Reordering Life

Inside Technology

Edited by Wiebe E. Bijker, W. Bernard Carlson, and Trevor Pinch

A list of the series appears at the back of the book.

Reordering Life

Knowledge and Control in the Genomics Revolution

Stephen Hilgartner

The MIT Press
Cambridge, Massachusetts
London, England

This book was set in Stone Sans and Stone Serif by Toppan Best-set Premedia Limited. Printed and bound in the United States of America.

Library of Congress Cataloging-in-Publication Data

Names: Hilgartner, Stephen, author.
Title: Reordering life : knowledge and control in the genomics revolution / Stephen Hilgartner.
Other titles: Inside technology.
Description: Cambridge, MA : The MIT Press, [2017] | Series: Inside technology | Includes bibliographical references and index.
Identifiers: LCCN 2016030923 | ISBN 9780262035866 (hardcover : alk. paper)
Subjects: | MESH: Human Genome Project. | Databases, Genetic--ethics | Genetic Research--legislation & jurisprudence | Intellectual Property
Classification: LCC QH447 | NLM QU 450 | DDC 572.8/629--dc23 LC record available at https://lccn.loc.gov/2016030923

10 9 8 7 6 5 4 3 2 1

To Kate

Contents

Acknowledgments

I could not have undertaken a project of this scope without the assistance and support of many people. I am especially indebted to my interlocutors in the world of genomics. Although the strictures of confidentiality prevent me from thanking them by name here, I greatly appreciate the generosity with which they welcomed me into their laboratories and meetings; gave freely of their precious time; shared their knowledge, opinions, hopes, and frustrations; tolerated intrusive questions; referred me to their colleagues; and assisted me in many other ways, large and small. Interacting with them was a pleasure and a privilege, and their excitement and enthusiasm was often contagious. I will always remain extremely grateful to them for so willingly sharing their worlds with me.

A number of colleagues and friends carefully read part or all of the manuscript, including Angie Boyce, Peter Dear, Michael Dennis, Jenny Gavacs, Rob Hagendijk, Jonathan Luskin, Ilil Navah-Benjamin, Nicole Nelson, Shobita Parthasarathy, Trevor Pinch, Judith Reppy, Kasia Tolwinski, and several anonymous reviewers. Their astute comments and criticisms improved the manuscript considerably.

Over the years many people contributed to this project through helpful conversations, challenging questions, and other forms of assistance, encouragement, and advice. I particularly thank Richard Bell, Carin Berkowitz, Charles Bosk, Sherry Brandt-Rauf, Alberto Cambrosio, Adele Clarke, Mehmet Ekinci, Ulrike Felt, Joan Fujimura, Jean-Paul Gaudillière, Tom Gieryn, Dave Guston, Linda Hogle, Tony Holtzman, Ben Hurlbut, Pierre-Benoit Joly, Peter Keating, Timothy Lenoir, Sabina Leonelli, Bruce Lewenstein, Javier Lezaun, Susan Lindee, Michael Lynch, Clark Miller, Yoshio Nukaga, Alain Pottage, Barbara Prainsack, Jenny Reardon, Hans-Jörg Rheinberger, Dirk Stemerding, Kaushik Sunder Rajan, Mariachiara Tallacchini, and Peter Weingart. It was a pleasure to work on the diagrams with Chris Cooley and Ranjit Singh, and I thank Chip Aquadro for his invaluable comments on several of them.

The faculty and graduate students of Cornell's Department of Science & Technology Studies created a stimulating intellectual environment, and the department's dedicated staff—especially our administrative manager Deb Van Galder—has earned my heartfelt appreciation. At MIT Press, I thank my editor, Katie Helke, for her enthusiasm and assistance, Virginia Crossman for her editorial advice, and Annie Barva, who copyedited the manuscript. Chris Hesselbein prepared the index.

Three people stand out for special thanks: David Rothman for his enthusiastic support of this research during its early stages; the late Dorothy Nelkin for helping me launch a career in social studies of science; and Sheila Jasanoff for her inspiring intellectual vision and many forms of guidance over the years.

No written acknowledgment can adequately express the depth of my gratitude to my family. I thank my mother for her love and support, and for her loving presence in the lives of my children. Mel, her late husband, is sorely missed. My father's enthusiasm for science was an important influence during my early years. My children, Nathan, Kevin, and Erin, continually inspire me, and I have known no greater joy than being their dad. And then there is Kate. The joining of our lives many years ago was a wonderful and transformative change. I cannot thank her enough for her insightful comments on my work, for her wisdom and support, and most of all for being the love of my life.

* * *

The research reported here is based on work supported in part by grants from the National Institutes of Health (NCHGR ELSI Program) and the National Science Foundation (Grants No. 0083414 and 035200). The findings and interpretations expressed in this book do not necessarily reflect the views of the National Institutes of Health or the National Science Foundation.

List of Abbreviations

ABI   Applied Biosystems Incorporated
BAC   bacterial artificial chromosome
cDNA   complementary DNA
CGR   Whitehead/MIT Center for Genome Research
CRP   community-resource project
DOE   US Department of Energy
ELSI   Ethical, Legal, and Social Implications
EMBL   European Molecular Biology Laboratory
EST   expressed sequence tag
HGMP   Human Gene Mapping Program
HGP   Human Genome Project
HGS   Human Genome Sciences
HUGO   Human Genome Organization
ICRF   Imperial Cancer Research Fund
IPO   initial public offering
LASL   Los Alamos Sequence Library
MIT   Massachusetts Institute of Technology
MRC   Medical Research Council
NAR   Nucleic Acids Research
NCBI   National Center for Biotechnology Information
NCHGR   National Center for Human Genome Research
NHGRI   National Human Genome Research Institute
NIH   National Institutes of Health
NRC   National Research Council
OBER   Office of Biological and Environmental Research
PCR   polymerase chain reaction
PTO   US Patent and Trademark Office

R&D	research and development
RFA	Request for Applications
RLS	Reference Library System
SCW	Single Chromosome Workshop
SNP	single nucleotide polymorphism
STS	sequence-tagged site
TIGR	The Institute for Genome Research
UJAD	unpublished in journals and available in databases
WGS	whole-genome shotgun
YAC	yeast artificial chromosome

1   Introduction

Understanding the dynamics of scientific and social change is an urgent challenge for the twenty-first-century social sciences. Consider the "genomics revolution." In the mid-1980s, a scientific vanguard emerged with the explicit aim of revolutionizing the biological sciences and medicine. These scientists sought to make *genomes*—the totality of an organism's inherited DNA—into tractable objects of analysis, finding all the genes "once and for all" and "creating a new biology for the twenty-first century."[1] Their vision of transforming the sciences of life, manifested most dramatically in calls for a Human Genome Project, soon captured imaginations, winning a $3 billion commitment from the US Congress. By June 2000, when US president Bill Clinton and UK prime minister Tony Blair held a joint news conference to announce that a "first survey" of the entire sequence of the human genome had been completed, genomic knowledge and technology had become indispensable to biological research and to the biotechnology and pharmaceutical industries.[2] The rise of genomics, most observers agreed, portended even greater change to come, not only in the new life sciences but also in the lives of ordinary citizens.

During the 1990s, biologists and others increasingly integrated genomic perspectives into ways of doing and funding research, producing significant change both within and beyond the scientific community. New factory-style laboratories took shape, greatly accelerating data production but fitting poorly with established ways to organize work and careers in molecular biology. The difficulty of working on a genomic scale made translaboratory cooperation more desirable, leading to conflicting visions of how best to orchestrate it. The quantity of data in the genome databases used by the scientific community continued to grow exponentially, destabilizing the existing systems for collecting and circulating data and challenging long-standing practices of scientific publication. Scientists and entrepreneurs reimagined genomic information as a form of capital, inspiring the formation of a new wave of companies and provoking intense debate about the ownership of the human genome. The prospect of a revolution in medicine inspired hope but also raised ethical, legal, and social questions. Such changes did not emerge from an orderly scientific debate or a straightforward policy-making process but through dynamic—and often contentious—negotiations among many actors in many institutional locations.

This book examines the "genomics revolution," treating it as an empirical site for investigating transformative scientific and social change. My study concentrates on what is often seen as the scientific and symbolic centerpiece of genomics: the Human Genome Project (HGP), an international effort that ranks among the most celebrated scientific achievements of the late twentieth century. Bringing genomics into being during the HGP was simultaneously an epistemic, material, and political process, and I examine how problems of governance arose in concert with new genomic knowledge and technology. Although some of these problems were addressed through official policy, others were settled in ad hoc forums and informal encounters. Controversial issues emerged at many levels, ranging from the individual laboratory to the field of genome research, to the wider worlds of science, industry, and society. Scientists, along with many other actors, contested a variety of questions: What forms of accountability should govern the production and use of genomic knowledge? Who should own genomic knowledge, and what should "ownership" mean? Under what conditions should researchers publish or withhold data? How should translaboratory collaborations be structured? And how should researchers interact with the news media and respond to societal concerns?

Examining how such questions were contested and resolved allows me to investigate how *new forms of knowledge* and *new modes of control* took shape as genomics emerged. From this point of view, the genomics revolution entailed not only change in knowledge and technology but also change in "knowledge-control regimes"—structures that allocate entitlements and burdens pertaining to knowledge. I will elaborate this concept more fully below. For now, suffice it to say that in a manner broadly analogous to the law, knowledge-control regimes constitute modes of control that apply to specific actors, entities, and jurisdictions—although they vary in the extent to which they are formally codified. In the biological sciences, computing and information technology, and many other sociotechnical domains, the processes that reconfigure knowledge-control regimes are an important aspect of the politics of contemporary societies.

This study is an empirical one, and I collected data using ethnographic and interview methods. Participant observation and interviewing took place in laboratories, scientific meetings, government agencies, biotechnology companies, and forums addressing ethical, legal, and social issues. Most of my fieldwork was conducted in the United States, and my account focuses on that country, but I also did some field research and interviews in France, Germany, and the United Kingdom. Because my goal was to investigate change rather than to characterize a relatively stable area of research, it was necessary for me to follow the action over time. The field research underlying this book began in 1988, two years before the formal launch of the genome project, and ended (with the exception of some follow-up fieldwork) in 2003, the year when the HGP was officially completed.[3] I introduce the field of genomics and the HGP in chapter 2, but my first order of business is to outline the theoretical framework and research methods underlying this study.

Transformative Scientific Change

To analyze the dynamics of change in knowledge and governance during the HGP, this book focuses on the contestation and reallocation of *control* during the production of genomic knowledge and technology. Put otherwise, I use this revolutionary moment as a convenient place to examine how modes of control were reconfigured as knowledge changed (and vice versa). As new forms of knowledge took shape, what adjustments were made in the lawlike regimes that governed biological research? How—in the laboratory and beyond—did actors attempt to use, resist, or change extant regimes and practices? How did they seek to establish and justify (re)allocations of rights, duties, privileges, and powers? Which regimes and practices proved durable, which changed, and why? And what implications did the process have for the distribution of authority, wealth, power, and the capacity to shape the future?

The existing literature on transformative scientific change offers some useful starting points for addressing such questions. Consider three famous examples, beginning with Thomas S. Kuhn's *The Structure of Scientific Revolutions*. For Kuhn, scientific revolutions overturn the extant paradigms that govern the research process, transforming scientists' conceptual, theoretical, instrumental, and methodological commitments. Knowledge does not simply accumulate but is fundamentally reconstructed to fit the new worldview.[4] What constitutes a worthwhile research question, what counts as adequate evidence, and which scientific skills matter may all be redefined. Revolutions thus reorder both what is known and the community of knowers. In this sense, Kuhn treats knowledge and social order as forming a single system that revolution transforms—an approach I build on here. Yet the social order that he describes is internal to the scientific community, set apart from the wider society.[5] Kuhn's revolutions occur *within* science, and he is relatively uninterested in how change in scientific paradigms connects to broader social change.[6] After a revolution brings a new paradigm into being, he writes, scientists may inhabit a new world, but outside the laboratory everyday affairs usually continue much as they did before.[7]

Ludwik Fleck's book *The Genesis and Development of a Scientific Fact*, published in 1935, presents another important analysis of transformative scientific change. Key aspects of Fleck's account anticipated *The Structure of Scientific Revolutions*, offering what is ultimately a richer sociology of knowledge.[8] Fleck's concept of a "thought style"—a closed, stylized system that constrains thinking and the interpretation of evidence—resembles Kuhn's paradigms. His concept of a "thought collective" composed of individuals who share a thought style also encompasses what Kuhn would call a scientific community, but Fleck's concept is broader.[9] All individuals belong to many thought collectives, which are not associated exclusively with science but also with commerce, the military, the arts, politics, religion, fashion, and so forth. In each domain, some thought collectives are relatively specialized, whereas others are less so. These thought collectives are tied together through "multiple intersections and interrelations ... both in space and time." Moreover, scientific ideas are embedded in society; culturally available "proto-ideas" promote their emergence and acceptance. Fleck also argues that new knowledge does not simply diffuse unchanged. It inevitably undergoes a "stylized remodeling" as it moves from the esoteric "vanguard" of a scientific field to increasingly exoteric thought collectives that incorporate it into their ideas and practices.[10] Fleck thus describes a dynamic process in which, as Ilana Löwy explains, "expert knowledge is influenced by popular knowledge, and then influences it in turn."[11] Transformative scientific change unfolds through interactions within and among thought collectives that are thoroughly embedded in modern societies.[12]

Finally, consider Bruno Latour's study *The Pasteurization of France*. Latour rejects the idea that the revolution attributed to Louis Pasteur can be analyzed by any approach that takes for granted an a priori distinction between "science" and "society."[13] The world is composed of hybrids of nature and culture, so understanding this "revolution" requires avoiding all distinctions between the content and context of science, between the natural and the social, even between human and nonhuman agency. Rather than attributing social change to scientific change or vice versa, Latour investigates the reconfiguration of networks of associations connecting heterogeneous actors while rejecting such distinctions as human/nonhuman or natural/ social. His method involves tracing the process through which new actors— such as Pasteur's "microbes"—are constituted as entities, circulated in texts, and built into material practices in medicine and hygiene. For Latour, this process of network building constitutes new "scientific knowledge" and new "social groups," but neither the knowledge nor the groups *cause* change; they are both the *outcome* of the reconfiguration of networks. Latour thus develops a metaphysical and methodological perspective on transformative change that treats "nature" and "society" as being constituted through a single process of reconfiguring networks.[14]

Each of these perspectives provides important insights about the nature of transformative scientific change. Yet neither Kuhn nor Fleck devote sustained attention to questions about the changes in the allocation of entitlements and burdens or questions of governance. Kuhn focuses narrowly on a scientific community that is strangely detached from the rest of society. Fleck's examination of intersecting thought collectives addresses this problem, but his project—to provide a historically and socially situated account of epistemic change—is not concerned with the distribution of power or the machinery of governance. Latour focuses on the emergence of new forms of power and offers a deep view of how knowledge and social order change together. His important insights about hybridity and heterogeneity, and his attention to the role of material entities in establishing order, are particularly helpful to my analysis. But his emphasis on metaphysical matters directs inquiry away from the critical and normative questions of greatest concern to people interested in how extant allocations of authority, wealth, and power are reordered during periods of transformative change in knowledge.[15]

The framework that Sheila Jasanoff terms "interactional co-production" offers a more promising theoretical point of departure for addressing such questions.[16] Jasanoff's approach builds on Steven Shapin and Simon Schaffer's *Leviathan and the Air-Pump*, extending its theoretical reach using comparative analysis of law and politics.[17] She proceeds from three main assumptions: (1) that ways of knowing the world and ways of ordering social relations are deeply interconnected, forming a single mode of ordering knowledge and society; (2) that established orders change through processes in which knowledge and social relations are mutually adjusted; and (3) that analyzing these processes requires being symmetrically attentive to changes in knowledge and governance. Her theoretical framework emphasizes such questions as how institutions, discourses, identities, constitutions, and imaginaries shape modes of decision making and guide public reason in specific societies.[18] Jasanoff's analytic perspective is well suited to exploring such politically significant questions as: During periods of transformative scientific change, how do reconfigurations of knowledge and lawlike modes of control take shape? And with what effects on the allocation of power, wealth, and the capacity to participate in processes that influence the future of sociotechnical change?

This book addresses these and similar questions by examining the coproduction of genomic knowledge (including technology) and the regimes and practices of control during the period of the HGP. In contrast to Jasanoff, whose research usually takes such sites as regulatory agencies, courts, and ethics commissions as empirical starting points, this study takes a specific scientific community as its empirical focus, following the activities of the genomics community in the laboratory and many other sites. As the vision of a new genomic knowledge was fleshed out, how did existing social orders integrate genomics into their practices? What changes took place in the lawlike regimes through which control was secured, contested, allocated, resisted, justified, and reconfigured? What possibilities were realized, which roads were not taken, and how can we account for these outcomes?

Three forms of control are particularly relevant to my investigation:

1. *Control over objects.* One way that control is allocated among agents is by managing the social, spatial, and temporal distribution of "knowledge objects"—entities or things that contain or constitute knowledge. Knowledge objects include not only scientific results, data, and information but also preliminary findings and plans, inscriptions and biomaterials, instruments and skilled personnel, techniques and software, as well as rumor, speculation, and scuttlebutt. An inclusive definition is needed because a wide range of materials, inscriptions, and devices are involved in the production of knowledge and control.[19] In examining genomics, I pay close attention to the distribution and contestation of control over knowledge objects and examine how change in knowledge objects is implicated in changes in control.

2. *Control over jurisdictions.* Control is also allocated by constituting "jurisdictions"—bounded spaces (physical, sociopolitical, and discursive) onto which agents and their capacities are mapped.[20] Because actors often contest jurisdictions, particularly during periods of change, the rhetorical moves that Thomas F. Gieryn calls "boundary work" are often central to struggles over control.[21] For example, classifying a contentious question as a matter of "science" or a matter of "policy" may have a strong bearing on who is entitled to decide the issue and how.[22] In examining genomics, I pay close attention to the (re)structuring of jurisdictions and to jurisdictional contests and boundary work.[23]

3. *Control of relationships.* Taking a relational view of control, I systematically investigate regimes and practices that allocate entitlements and burdens among agents. Like the legal relationships identified by Wesley Newcomb Hohfeld,[24] "control relationships" structure the terms of interaction among agents. If a control relationship grants one agent a specific entitlement (e.g., to prohibit trespassing on her property), it exists with respect to another agent (or the class of similarly situated agents) who bears a correlative burden (e.g., not to enter without permission). This study examines how control relationships formatted the roles and conduct of agents, how actors sometimes resisted control, and how control relationships changed during the action in the rise of genomics. For example, I analyze how science policy makers attempted to introduce new forms of accountability to control genome research and I consider ways that researchers adapted to—and sometimes resisted—such controls.

These three forms of control—of objects, of jurisdictions, and of relationships—do not operate separately; they are best thought of as aspects of a single, dynamic process through which specific configurations of knowledge and control are made, reproduced, and changed. I investigate this process in a variety of sites, including face-to-face interactions among scientists, exchanges among laboratories, policies to guide research programs, collaborations among companies and between companies and scientists, and attempts to shape the interpretive narratives about genomics displayed to publics.

Knowledge-Control Regimes

To analyze the rise of genomics, this book employs a theoretical framework centered on what I call "knowledge-control regimes." The term *regime* is used widely in the social sciences in a variety of contexts. Most usage shares the idea that a regime imposes order on a domain or activity, typically through some combination of formal rules, informal norms, material means, and discursive framings.[25] As a first step in introducing the concept of a knowledge-control regime, let me offer some examples: military classification, trade secrecy, the regulation of confidential human-subject information, and the legal instruments that formally codify intellectual property.[26] The concept, however, is not limited to regimes that restrict the flow of knowledge objects; it also encompasses those intended to accomplish the reverse, including the regime of authorship and publication in scientific journals and such "open-source" regimes as the General Public License, Creative Commons, and the BioBricks Initiative.[27] Genomics companies' business models—which constitute particular modes of control over knowledge—also qualify as knowledge-control regimes. So do the procedures that science advisory committees use to regulate access to information while they prepare written reports.[28] A "moral economy" of shared expectations about resource exchange also falls within this category, as do the public-relations practices that corporations and governments use to shape news coverage.[29] Even ad hoc, short-term arrangements, such as informal agreements between laboratories to "collaborate," fall within its scope.[30]

As these examples—*all* of which are relevant to genomics—suggest, knowledge-control regimes operate on a wide range of scales and are formally codified to differing degrees, relying on many kinds of legal and quasi-legal mechanisms as well as on collectively understood templates for engaging in "guided doings."[31] Knowledge-control regimes also vary in their degree of institutionalization, and they regulate domains of activity that need not correspond with the external or internal boundaries of formal organizations. They facilitate many goals, such as allocating epistemic authority, distributing credit, creating property, spreading knowledge, maintaining privacy, ensuring quality, protecting national security, saving face, constructing professional jurisdictions, and shaping future histories. What unifies this broad category is (1) the central role such regimes play in regulating the production and use of knowledge, and (2) their lawlike structure, which constitutes specific means of controlling knowledge objects, disciplining actors, and bringing order to specific jurisdictions.

A *knowledge-control regime* can be defined as a sociotechnical arrangement that constitutes categories of agents, spaces, objects, and relationships among them in a manner that allocates entitlements and burdens pertaining to knowledge. By *agents*, I mean constructs to which agency is attributed—such as the "author," the "inventor," the "company," and the "human research subject." These constructs serve as framing devices to describe generic arrangements (in which types of actors are positioned) or to conceptually package specific situations (in which identifiable actors are implicated).[32] Regimes also constitute *spaces*, such constructs as "the individual's body," the "scientist's laboratory," or "the published literature," in which knowledge objects may be found and over which agents may enjoy various forms of control. Finally, knowledge-control regimes constitute *control relationships* that allocate entitlements and burdens among agents.

These control relationships might be described as "rights," but there is analytic value in distinguishing among the variety of relationships that the language of rights often sweeps into the same undifferentiated bin. Like Hohfeld, I reserve the term *rights* for those relationships that impose correlative *duties* on others. These true rights, he contends, should be distinguished from what he terms privileges, powers, and immunities, each of which also has a corresponding correlative.[33] Thus, a *privilege* imposes no duties on others but positions other parties as possessing *no right* to interfere with its execution. Thus, you have a privilege, not a right, to sell your favorite hat because you are at liberty to sell it, but no one has a duty to buy it. A *power* entails a correlative *liability* (that is, obligation), and an *immunity* constitutes a correlative *disability*. The point is not to adopt Hohfeld's typology wholesale.[34] It is rather (1) to underline how knowledge-control regimes position agents in mutual but asymmetric correlative relationships with one another, and (2) to emphasize the variety of control relationships that specify which agents are endowed with what entitlements and burdens as they pertain to other agents and to control over spaces, objects, and actions.

Control relationships are typically constituted by hybrids of material, discursive, and social means, and specific relationships can be adjusted, transferred, or stripped away using any of these means. No scientist, for example, has the privilege of piercing the human subject's skin with a needle to extract "interesting" tissue without the subject's explicit permission. The tissue is neatly contained, at least in the imaginary of bioethical regulation, not only inside the physical boundaries of the subject's body but also within a regime of human subject protections and, more broadly, a regime of individual rights that impose a duty to obtain consent.[35] Yet the human subject has the option of granting freedom of action to the researcher. With the legal instrumentation of informed consent and contract law, the privilege of controlling future use of the material can be extracted from the subject along with the tissue sample and transferred to the scientist. Residual encumbrances, such as a duty to keep the source's identity confidential, may also be legally constituted, attached to the material, and transferred along with the material to future users. As this example suggests, encumbrances may be both added and removed using legal devices (e.g., material-transfer agreements, nondisclosure agreements, and military classification) that allow targeted, partial, or limited circulation and use. The patent regime, which requires unconstrained disclosure of descriptions of inventions while granting monopolistic encumbrances on their material use, offers another example.

Many knowledge-control regimes, as the example of patents suggests, constitute forms of property. Social scientists have broadened the concept of property considerably in recent decades.[36] Elinor Ostrom and her colleagues, for example, have extended the traditional division of property into private goods, state-owned property, and public goods by adding the category of "common-pool resources."[37] The rise of free and open-source software, along with such innovations as Creative Commons, has further extended our conceptions of property to include these new "open" forms.[38] Research in science and technology studies has rendered visible the work required to make knowledge public, showing that no intrinsic characteristics of knowledge require it to be a public good and demonstrating that no form of knowledge is impossible to commodify.[39] The concept of appropriation practices has thus come to encompass means of creating a wide range of "public" as well as "private" properties.[40]

However, not all knowledge-control regimes neatly map onto even these expanded concepts of property and appropriation. Knowledge-control regimes extend to a variety of metaphoric forms of "ownership," such as the jurisdiction of a profession, the molecular biologist's right to control "his" or "her" laboratory, the epistemic/political authority to make certain kinds of decisions, and the individual's zone of privacy. Jurisdictions also include what Joseph Gusfield calls "political responsibility," such as the sometimes unwelcome obligation to "do something" about a public problem.[41] Knowledge-control regimes therefore pertain to much more than establishing the boundary, so important in capitalism, between knowledge that is "public" and knowledge that is "private" (in the sense of tradable).[42] They are not narrowly "economic" in character but are arrangements that constitute and allocate an amalgam of epistemic and sociopolitical authority.[43]

Governing Frames

The conceptual structure of a knowledge-control regime is encoded in what I call its "governing frame." I use the term *frame* here, following Erving Goffman, to designate an organized set of schemata that individuals and collectives use to interpret situations or activities as being instances of a particular kind of event or deed. People use frames to form a view about the question "What is it that's going on here?"[44] In everyday life, an individual uses frames to recognize a situation as an instance of a type of activity. Thus, the commonplace frame "paying at a store" is a conceptual resource that clerk and customer alike use to formulate expectations and guide their actions as the encounter unfolds. Frames vary in their degree of organization. Some, Goffman writes, are "neatly presentable as a system of entities, postulates, and rules; others—indeed, most others—appear to have no apparent articulated shape, providing only a lore of understanding, an approach, a perspective." Whether tightly or loosely organized, each frame "allows its user to locate, perceive, identify, and label a seemingly infinite number of concrete occurrences defined in its terms."[45] Beyond mere labeling, frames provide organized expectations and templates that actors use to guide normal or appropriate action. Frames are sustained by the ongoing practices of the societies in which they are embedded. They are typically inscribed not only in the memories of individuals but also in hybrids of the social, the discursive, and the material (for instance, think of law and cash registers in the "paying at a store" example).

Like the familiar frames of everyday life, the governing frame of a knowledge-control regime is an organized set of schemata that provides a template that actors employ to guide action and interpretation. Such frames play an important—or, better, constitutional—role in the operation of knowledge-control regimes.[46] The adjective *constitutional* is meant to invoke the metaphor of the modern state to underline how a regime's governing frame provides a stylized depiction of the most fundamental agents, things, processes, and rules that—at least in official representations—are said to guide its operation. The governing frame of a democratic state, perhaps inscribed in a written document such as the US Constitution or the German Basic Law, typically defines such key agents as "the executive," "the legislature," "the judiciary," and "the citizens," specifying their official roles and outlining their prescribed modes of interaction. Such a frame provides a schematic representation of its key elements and the relationships among them. It identifies physical spaces (such as the United States and its territories) and metaphoric ones (such as the writing of legislation) over which these agents are granted jurisdictional authority. It highlights a set of objects (such as laws or taxes) and actions (such as speaking, voting, appointing, or impeaching) that are salient elements of the frame. In short, such a frame promotes an official viewpoint that endows agents with specific entitlements and burdens pertaining to other agents or to control over spaces, objects, and actions.

Because constitutions pertain to the basic elements of a regime, they define entities—whether agents, things, spaces, or actions—at a rather generic level. Regarding agents, for example, the governing frame of a modern state generally refers to stylized *characters*, such as "the president," not to specific *actors*, such as the human beings George W. Bush or Barack Obama, whom the rough and tumble of historical events happened to cast in the presidential role.[47] Similarly, key spaces, objects, and actions are also presented in generic terms. As a schematic account of how things are supposed to work, a governing frame provides a stylized map of categories and processes, not a complete script, and it inevitably leaves room for actors to produce conflicting interpretations of how the regime does or should operate. Actors often contest threshold issues, argue about whether agents are overstepping their authority, struggle about how to classify novel situations, and sometimes question the legitimacy of the frame itself. Nevertheless, the governing frame typically supplies the ontology within which such disputes are conducted. To be sure, the governing frames of democratic states inevitably change over time, sometimes gradually and sometimes abruptly.[48] Yet these frames tend to display sufficient stability to play an ongoing role in the process of governing, even when the action deviates sharply from the simplified paths that the frames depict.

The governing frame of a knowledge-control regime has similar properties, and analyzing its structure offers a means to identify the core features of the kinds of knowledge and forms of control it instantiates. Consider, for example, the familiar regime that regulates publication in scientific journals, evaluating manuscripts and constituting the scientific literature (referred to throughout the book as the "scientific-publication regime" or the "journal regime" or "publication regime" for short). The governing frame of the publication regime defines key agents (authors, reviewers, readers) and roles (writing, reviewing, citing) while choreographing the flow of scientific texts through three jurisdictional spaces (unpublished, under review, and published) and specifying the control relationships operative in each. It circumscribes agents' entitlements and burdens as it formats their legitimate courses of action. The scientist who seeks to publish a manuscript cannot expect to engage with a journal in just any arbitrary way: to enter into the regime, she must allow it to cast her as a specific type of predefined character—namely, a "manuscript author" whose work will be subjected to peer review and possibly earn a place in the published literature. In more than one way, sending a paper to a journal is an act of submission!

The world of journals, of course, is by no means uniform, and owing to this variation we might wonder if what I have called its governing frame is nothing more than a misleading idealization. Like the official constitution of a state, which we confuse with its actual mode of governance at our peril, the governing frame of the publication regime clearly offers a grossly inadequate description of the ongoing practices that instantiate either the scientific-publication system as a whole or any individual journal. Key elements of the frame, such as "the scientific literature," are (on one level) semiotic entities that provide a convenient gloss on things that—in the totality of their full sociotechnical glory—are stunningly complex. Even fifty years ago "the scientific literature" consisted of countless papers in countless journals in countless libraries. Today, the basic question "Where can one find the scientific literature?" leads to an impossible tangle of geographic and virtual locations. But to argue on this basis that the semiotic object "the scientific literature" is irrelevant to the operation of journals would be to misconstrue the situation completely, for this object is a crucial *part* of the sociotechnical arrangements that produce, circulate, and store scientific texts.

By simplifying the infinite particularity of sociotechnical networks into a workable set of entities, relationships, and expectations, a governing frame provides a generic map of the most basic agents, objects, spaces, and processes underlying a regime. That these structures are generic is precisely what makes them so important to holding a regime together, for this fact allows a frame to retain its fundamental structure even as actors— individually and collectively—apply it to a wide range of more specific circumstances. The governing frame of the publication regime thus provides a generic structure that underwrites the (relatively) standardized sociocultural form of "the" scientific journal, allowing shared expectations and rules of interaction to permit coordination. This frame is thoroughly institutionalized in contemporary research systems and draws moral force from justificatory discourses about the progress of science and the need to reward discovery.

None of this is to say that governing frames are static; they often undergo minor and sometimes major adjustments, especially during times of rapid change—such as the period in which genomics emerged during the HGP. In examining problems of governance that arose as new genomic knowledge took shape, I pay close attention to the governing frames of knowledge-control regimes. What kinds of objects, spaces, and agents did these frames define, and how were jurisdictions bounded? What kinds of rights, privileges, powers, and other control relationships did they instantiate? What forms of accountability did they seek to create, channel, and check? And what changes in these governing frames occurred and why?

Multiple Regimes

Many knowledge-control regimes are relevant to the world of genomics, and the existence of multiple regimes—typically interconnected, overlapping, or nested within each other—raises a number of issues. For one thing, the core concepts of well-institutionalized knowledge-control regimes, such as scientific publication, are woven into many other regimes, such as those governing the appointment of university faculty members. Regimes thus can interconnect in ways that enhance (or sometimes reduce) their stability and strength. For another thing, actors frequently guide entities across regime boundaries, reformatting those entities in ways that transform them significantly. Transfers among regimes thus change knowledge objects as they put new control relationships and identities into effect.

Through this process, a single knowledge object can differentiate into a variety of forms.[49] Thus, a line of scientific work conducted in Leroy Hood's laboratory was packaged into both a *Nature* paper titled "Automated Florescent DNA Sequencing" and a closely related piece of intellectual property, US Patent No. 5,171,534—which became one of the most important patents pertaining to genome technology of the past three decades.[50] Even what appears to be the "same" object can be situated in more than one regime, taking on a variety of meanings and characteristics as different governing frames (re)cast it in their terms. The scientific-publication regime may frame the *Nature* paper as an object in the open-to-any-and-all space of "the scientific literature," but this does not stop Nature Publications Incorporated from charging *Nature* nonsubscribers $32 for an electronic copy—a fact that highlights the notable gap between the de jure entitlements of the imaginary of the scientific-publication regime and the de facto entitlements in a world where scientific papers are often copyrighted commodities. In an important sense, the *Nature* paper is not the same knowledge object when viewed from the perspective of the scientific-publication regime and the copyright regime—a point not lost on the proponents of open-access publication.[51]

The complex topology of knowledge-control regimes may also raise jurisdictional questions about which control relationships are (or should be) operative in a given situation. Much of the time, knowledge-control regimes are not neatly laid out like the nations of a politically stable part of the world, where clear borders create distinct jurisdictions. Even along the most friendly of national borders, the crossing of people and objects may be problematic, and contentious jurisdictional questions sometimes arise. At the edges of technological change, jurisdictional questions may more closely resemble those found in geographic regions where the boundaries, legitimacy, and very existence of states are contested (think of the Middle East in 2016). Moreover, when jurisdictions arguably overlap, ambiguities arise about which regime should defer to which.[52] In examining genome research, I consider how conflicts that emerged at points of jurisdictional contact were negotiated and resolved.

Agency and Action

So far, in looking at the structure of knowledge-control regimes, I have considered the stylized characters that they constitute, paying less attention to the specific *actors*—humans, organizations, and groups—who engage with these regimes.[53] Such actors need not simply surrender their agency to the circumscribed roles into which regimes cast them; they can resist control or seek to expand their freedom of action, and they use a variety of practices to construct and exploit wiggle room. Knowledge-control regimes do not *determine* action but structure the categories, routines, identities, rules, expectations, menus of options, and other resources that actors draw on to interpret situations and guide action. Put otherwise, actors may try to use these resources when they act, or they may try to resist, stretch, ignore, or transform them.

The concept of a knowledge-control regime developed here in no way rests on a model that sees control as eliminating human and organizational actors' agency. In contrast to Mertonian sociology of science—with its vision of internalized values, strict policing, sanctions, and an affective commitment to institutionalized norms that tightly control conduct—the account advanced here emphasizes schema, routines, and strategic action on a shifting field of opportunities, constraints, and interpretive work.[54] This is not to say that policing and the threat of sanctions play no role in underwriting knowledge-control regimes; they clearly do. But rather than imagining conduct to be controlled by clearly defined, binding norms and sanctions that strip away agency, the perspective advanced here sees these regimes as constituting frames and modes of action through which—and against which—actors can exert agency.[55]

As we will see in the chapters ahead, actors often enjoy considerable room for maneuvering as they engage with knowledge-control regimes.

The ways in which they create and exploit maneuvering room are too numerous to capture in a simple typology. However, it is worth noting some salient ways through which they accomplish these things:[56]

1. Although knowledge-control regimes may prepackage a limited set of modes of engagement, actors often can choose whether and when to enter into one or more of the available options. In their interactions with the publication regime, for example, genome researchers (like other scientists) regularly faced the question of whether and when a scientific result was ready to "submit" to the regime's form of governance—in other words, whether to send out a manuscript.

2. Where multiple knowledge-control regimes are relevant, actors may strategically select from a menu of options, perhaps by choosing among regimes or by submitting to several regimes at once or by submitting to several regimes in succession. Genome researchers often chose both to file a patent application and to submit a scientific paper, and they sometimes argued about which was the appropriate course of action.

3. Actors may be able to launch, exploit, or provoke jurisdictional conflicts to advance their interests. Thus, the open-access movement has attacked journals' copyright policies, arguing that a key principle of the scientific-publication regime (once results are published, they can be freely used) should trump an equally central principle of the copyright regime (the copyright holder can set the terms of use). I analyze jurisdictional disputes pertaining to "publication" when genome databases joined journals as a means of dissemination.

4. Actors may be able to operate outside the regulatory reach of a knowledge-control regime. For example, they may be able to duplicate and circulate copyrighted papers or infringe on patents in ways that are undetectable or too trivial to merit a legal challenge.[57] Genome laboratories engaged in analogous practices.

5. Actors may be able to exploit the inevitable ambiguities in category membership judgments to create wiggle room. Journals seek to publish only "novel" findings, but what counts as sufficiently novel may be contested. Such forms of boundary work were common during the HGP.

6. Actors may be able to use gaps in the procedural rules of a knowledge-control regime to claim authority over decisions. The

scientific-publication regime promotes the expectation that authors will grant reasonable requests from qualified scientists for access to the data and materials underlying published papers. This norm applies to biological materials, such as DNA samples, that are crucial to building on published work. However, the regime does not specify who is empowered to make threshold judgments about such matters as what constitutes a reasonable request or whether good reasons for an exception exist. As we will see, genome scientists sometimes claimed the privilege of making such judgments themselves.

7. Actors sometimes can feign adherence to the rules of a knowledge-control regime while surreptitiously departing from the rules' letter and/or spirit.[58] Genome scientists sometimes worried or concluded that others were attempting to dupe them in this way, as discussed later, especially in chapter 3.

8. Finally, in principle, actors can challenge a knowledge-control regime directly—for example, by attempting to redefine its governing frame or by questioning its legitimacy or applicability to the matter at hand.

This list by no means exhausts the possibilities. It does, however, give a general sense of some of the ways in which actors construct and deploy maneuvering room as they engage with knowledge-control regimes. Although some of these eight examples express the perspective of agents who seek to enlarge their own scope of action by exploiting or expanding weaknesses in regimes, actors who seek to tighten control (over others or even over themselves) use a similar array of techniques. In examining the rise of genomics, I pay close attention to the strategic maneuvering by which actors struggled to tighten or loosen control or both.

Stability and Change

One final aspect of knowledge-control regimes remains to be discussed: the dynamic processes through which they are reproduced and change. Knowledge-control regimes are stabilized by complex imbrications of the material, the social, and the discursive, but even the most well-institutionalized regimes are susceptible to subtle—and, at least in principle, radical—change. Destabilization can result from shifts in the everyday practices and conceptual categories of the societies and groups in which regimes are embedded. Intentional action can also destabilize them because actors can at any time attempt to adjust, stretch, or even radically break a governing frame. Actors can also try to construct new knowledge-control regimes, usually by recombining elements of existing ones—as in the well-known example of the General Public License, which uses copyright and contract law to constitute a novel "free-software" regime.[59] Yet change in knowledge-control regimes rarely goes uncontested because those who benefit from the status quo tend to defend the regimes that entitle them to authority, wealth, and power. Thus, to the extent that a knowledge-control regime does in fact change, it is often only after a period of dynamic struggle.

Knowledge-control regimes are especially likely to be sites of intense controversy during periods of transformative scientific change. New paradigms and new technologies have the potential to perturb extant regimes, reopening normative choices and stimulating the introduction of new modes of control. Such periods thus offer an excellent opportunity to explore questions about the processes that constitute new knowledge-control regimes or that destabilize and restabilize existing ones.

The genomics revolution, although interesting in its own right, thus also offers an excellent site for exploring questions about change in knowledge-control regimes and the struggles that accompany it. Providing such an empirical exploration is the main goal of this book. Through what processes did knowledge-control regimes change during the genomics revolution? To what extent and in what ways were the governing frames of established regimes reshaped, and what accounts for the form of the settlements reached? Why did some new regimes fail to take hold but others were successfully constructed? When multiple regimes were interconnected, did change in one regime tend to destabilize "adjacent" ones, and why or why not? And can this study of the coproduction of knowledge and control in genomics provide insights relevant to other areas of transformative sociotechnical change?

Research Methods

The questions I have outlined can be answered only through empirical research using methods sensitive to meanings, values, and the complexities of social life. Accordingly, this study employs ethnographic observation, interviews, and analysis of documents. Unlike the classic ethnographies of scientific laboratories,[60] however, this study focuses not on a specific laboratory or epistemic culture but on an entire community—namely, the scientists, institutions, and others involved in the HGP. The practices used to amass the data undergirding my analysis are discussed in the appendix. Nevertheless, a few points about my research methods bear brief mention here.

In early 1988, I began looking for research sites for an ethnographic study of laboratory practices in molecular genetics. Reading about the nascent HGP convinced me that genome mapping and sequencing merited close attention. As a self-consciously "revolutionary" endeavor, the HGP aimed at effecting significant transformations in biological science, medicine, industry, and society. If successful, the genome project, I reasoned, would be implicated in notable epistemic, technological, and social changes. Were it to fail, its rise and fall would likewise surely be illuminating. The palpable enthusiasm of the genome scientists I met and their sense of the historical—and sometimes even mythic—significance of the HGP also intrigued me.

With the help of a colleague, I negotiated access in 1988 to a laboratory involved in genome mapping. The scientists introduced me to the laboratory much as they would a new graduate student; they trained me in their techniques and gave me a small research problem to solve. I hung around and even did a series of experiments, all the while observing laboratory practice, asking naive questions, and taking field notes. I was soon accompanying laboratory members to conferences and meeting new researchers, who welcomed me into their worlds and laboratories. My study of genomics expanded to include multiple laboratories and other sites. As I met new genome researchers, I joked about how I was like an anthropologist who was going hunting and fishing (for genes) with the natives—a remark that punned on the acronym FISH, short for "fluorescent in situ hybridization," a molecular genetics technique. The joke usefully distinguished me from a journalist, on the one hand, and from an ethicist, on the other, while indicating my goal of understanding their practices in social science terms.

In May 1989, a senior scientist invited me to drop in on that year's Cold Spring Harbor meeting on "Genome Mapping and Sequencing," the second in a series of annual meetings of the fledgling genomics community.[61] Another senior scientist recommended that I attend "Human Genome I," a conference designed to help launch the HGP. At such events, I made new contacts (including some from Europe) to arrange additional fieldwork at genome project laboratories and to set up audiorecorded interviews. Convinced that conducting a "prospective study" that would follow the field for a number of years would be an interesting methodological "experiment," I committed myself to doing long-term fieldwork in the broader community engaged in genome mapping and sequencing. After the scope of research expanded, it mostly involved interviewing, observation at scientific and policy meetings, and laboratory visits ranging from a day or two to about a week. In addition to US sites, I conducted research trips to the United Kingdom, Germany, and France in the early 1990s and to Germany again in 1998 and 2001. I continued to study the genomics community actively during the final years of the HGP and did some follow-up fieldwork after its official completion in 2003.

The genome laboratories studied were deeply involved in the HGP. My interlocutors included people engaged in genome mapping, sequencing, informatics research, database design and management, technology development, and the hunt for genes that cause disease. They included not only some of the leaders in the field but also postdoctoral researchers, graduate students, and technicians. I also interviewed scientists and managers at companies involved in genomics and genetic testing as well as some government officials. I assured my interlocutors that their names would not be revealed in publications. Accordingly, I have assigned pseudonyms to individuals (e.g., NATHAN, KEVIN, and ERIN) or describe them using brief identifying phrases (e.g., GENOME MAPPER) presented initially in small caps. I have assigned a few people more than one pseudonym because otherwise their unique biographies might make them easily identifiable. The names of such organizations as laboratories (MAPPING LABORATORY), companies (COMPANY ONE), and universities (EASTERN UNIVERSITY) are handled in a similar manner. Where sensitive issues are involved, I have (possibly!) included red herrings such as a misleading geographic location. The reader should be warned that EASTERN GENOME CENTER might be in California. In general, I was impressed with people's willingness to talk to me and with the freedom with which they described their work and their views to me. When people were less than forthcoming, I treated this silence as a form of data as well as a constraint. I am enormously grateful to my interlocutors and greatly appreciate their generosity and their telling me as much or as little as they did.

Participant observation at scientific meetings and science policy meetings was also central to this study. The Cold Spring Harbor meetings on "Genome Mapping and Sequencing," where HGP scientists gathered annually, were particularly important to my research, and I attended (parts of) these meetings as an observer in 1989–1995, 1998, 2000, and 2001.[62] Throughout the book, ethnographic vignettes derived from field notes appear in a sans serif font.

This book also draws on a wide range of published materials, such as the existing literature on the HGP, including work by scientists who participated in the project, science journalists, and the science policy analyst Robert Cook-Deegan.[63] In contrast to interviews and field notes, I cite published materials in the usual manner, providing the real names of individuals and organizations. At times, I use information from unpublished personal letters, memoranda, and the internal documents of various organizations. I treat such documents like interview excerpts, providing general information about sources in ways that do not compromise individual identities.

The Chapters Ahead

This book is organized into eight chapters. The order of the chapters is loosely chronological, but my aim is not to present a comprehensive history of the rise of genomics or of the HGP. Instead, I investigate the HGP as a means to develop a theoretically informed analysis of change in knowledge and knowledge-control regimes that is solidly grounded in an empirical study of a self-consciously revolutionary scientific community. To set the stage, chapter 2 introduces genomics and the vision of the scientific vanguard that launched the HGP, and outlines the sociotechnical challenges that the genome project faced. The chapter also provides an introduction to genome mapping and sequencing, and presents brief ethnographic sketches of several mapping and sequencing laboratories.

Chapter 3 has two central aims. First, it elaborates an account of the regimes and practices implicated in securing knowledge and control in human genetics and genomics laboratories at the outset of the HGP (1988–1992). It describes how laboratories managed inbound and outbound transfers of knowledge, shows how laboratory practices can simultaneously shape both knowledge objects and control over them, and examines some complexities of multi-laboratory collaborations. Second, the chapter serves to establish a baseline picture of the regimes and practices during this early period against which we can contrast change during the HGP.

The next four chapters each look at the coproduction of knowledge and control from a different angle, while also presenting a loosely chronological account of the HGP. Chapter 4 focuses on regimes intended to govern research programs and the scientific communities that conduct them. At the outset of the HGP, members of the genomics vanguard were convinced that new ways to coordinate science would be needed to realize their goals, and scientists in the US and Europe constructed a variety of regimes for doing so. The chapter performs a comparative analysis of two of those regimes, each of which advanced a very different normative vision of how to orchestrate genome research. The chapter examines the development of each regime, systematically comparing the modes of control each sought to instantiate and the sociotechnical means through which each did so. I also discuss the resistance that each regime encountered. The chapter concludes by offering an explanation of why one of these regimes was far more successful than the other.

Next, chapter 5 takes a different tack, focusing on a set of important knowledge objects in order to explore how new objects and regimes took shape together. I selected these objects for scrutiny because they played an important role in debate about the goals of the HGP and in the rise of the first genomics companies specifically defined as such. Tracing how each object was developed, transformed, and entangled in struggles about the future of genome research enables me to examine the role of law, policy, and business plans in the process of shaping new knowledge objects and regimes. The chapter concludes by reflecting on which visions of change came to pass, which did not, and whether there is a pattern in the outcome.

Whereas chapter 5 traces the struggles surrounding a set of important objects, chapter 6 traces the rise and fall of a series of important regimes. This approach allows me to focus on the dynamics of regime change, paying attention to both the emergence of new knowledge-control regimes and the interactions between the new regimes and more established ones. To accomplish this, the chapter examines the destabilization and restabilization of regimes associated with genome databases from the early 1980s until 2003, analyzing ongoing efforts to (re)negotiate settlements among databases, journals, funding agencies, and sequencing laboratories. This approach allows me to empirically explore a process of mutual adjustment among regimes at points of jurisdictional contact.

Chapter 7 examines an episode of intense competition between two factions of the genomics vanguard, the HGP and Celera Genomics, a private company established in 1998 with the goal of sequencing the human genome before the HGP was completed. The parties contested many issues, but in my judgment the issue with the highest stakes concerned the different knowledge-control regimes that each faction sought to instantiate. The chapter pays particular attention to the efforts of each side to win hearts and minds, and I analyze the narratives of justification that each deployed. As this competition neared an endgame, the HGP was drawn to a close though a series of celebratory events, the first of which, presided over by President Clinton and Prime Minister Blair, recognized both the HGP and Celera for the achievement. The chapter concludes by examining Clinton's celebratory speech, the sociotechnical imaginaries that it expressed, and the production of a historical achievement as a public fact.

The conclusion, chapter 8, summarizes the findings of this study and reflects on its implications for understanding the coproduction of knowledge and control during transformative scientific change and for the politics of emerging technologies. And the appendix reviews my methodology and research practices.

2   Envisioning a Revolution

To set the stage for the rest of the book, this chapter introduces the genomics vanguard, the vision of genomics that its leaders hoped to realize, and the challenges that they expected to face at the outset of the HGP. It also provides background information on genomes and genome technology.

*   *   *

What sort of "revolution" did the members of the genomics vanguard hope to catalyze? The scare quotes around the term *revolution* reflect not only caution about the indiscriminant, promotional use of the term but also the doubts that contemporary scholars in historical and social studies of science harbor about the concept, which smacks of an implicit teleology of progress.[1] It is therefore worth clarifying what I mean by referring to the genomics "revolution."

In manifestly political domains, the term *revolution* generally signifies some form of upheaval resulting in regime change within a given polity, but its precise meaning remains problematic. How radical must change be to earn the designation? Do coups that replace one elite with another fall into the category, or are far-reaching transformations of social relations required?[2] Revolutionary periods also have ambiguous boundaries, as do the social movements that bring them into being. Even in hindsight, the precursors to political revolutions may be indistinguishable from ordinary conflict.[3]

Scientific and technological revolutions raise analogous problems. What counts as a truly revolutionary change? How can the beginnings and ends of scientific revolutions be separated from their antecedents and consequences? To complicate matters, people can isolate and recombine intersecting "revolutions" in many ways. The "genomics revolution" can be framed as part of a wider "biological revolution," the "Internet revolution" as part of a wider "computer revolution," and both of these wider revolutions as part of a still more encompassing "information revolution." Furthermore, accounts of revolutions—whether scientific or political or both—are rarely innocent. Proclaiming a revolution is a performative speech act, one that does not simply describe the world but *acts* on it, often energizing both unbridled enthusiasm and determined opposition.[4] In science, in politics, and in the large and expanding hybrid domain that blends the two, revolution is a contested concept.[5]

In my view, the rise of genomics easily looks like a sufficiently far-reaching change to earn the title "revolution." But skeptical readers should note that from a methodological standpoint the question of whether the term *revolution* properly applies to genomics is not terribly important to this study. My central goal—investigating the coproduction of knowledge and control—does not depend on whether genomics arose through a "revolutionary" or an "evolutionary" process. Either way, genome research provides ample empirical material for investigating how knowledge and control are reshaped during a period of transformative scientific change. Identifying precise beginnings and endpoints of the period of transformative change is similarly unnecessary. All that is required is to study a slice of significant change and carefully analyze the process through which it occurred.

Nevertheless, in one crucially important respect the concept of "revolution" is absolutely necessary for understanding genomics during the HGP: *the scientists who launched the HGP often self-consciously cast themselves as leaders of a scientific revolution.*[6] Genome scientists compared the HGP to the Apollo moon shot, the search for the Holy Grail, the creation of the periodic table, and a journey to the center of biology. Even when they described genomic technology as mundane "infrastructure," they made it clear that this infrastructure would enable unprecedented discoveries. Members of the vanguard framed their project as one that would transform biological research, medicine, biotechnology, and, in many ways, life itself, bringing extraordinary improvements to the human condition. They expected the HGP to fundamentally alter both science and society, and they believed that it would long be remembered as a major event in human history. Belief in the transformative potential of genomics was an extremely important collective commitment of the leaders of the field, who fashioned themselves as standing at the forefront of fundamental change. This revolutionary identity gave them energy and a sense of urgency, encouraging them to be "audacious" and to challenge established ways of thinking and doing.[7] Similar self-understandings are sometimes found among other scientific and technological vanguards, whose visions often celebrate disrupting the status quo and replacing the old with a radically different new.[8]

The Genomics Vanguard

During the second half of the 1980s, the vanguard that proposed what became the HGP emerged. This group is an instance of what I have elsewhere called a "sociotechnical vanguard"—that is, a relatively small collective that formulates and acts intentionally to realize an edgy "vanguard vision" of a future attainable through science and technology. The term refers to visions that have not (yet) been widely accepted, and many of them never are.[9] "Vanguard visions" should be distinguished from what Sheila Jasanoff and Sang-Hyun Kim call "sociotechnical imaginaries"— the broadly shared, institutionally stabilized, and publicly performed aspirations that larger collectivities, such as nations, imagine as achievable through science and technology.[10] Sociotechnical imaginaries typically have relatively *longue durée* histories and exhibit greater stability than do the numerous and often fleeting visions developed by self-proclaimed sociotechnical vanguards. Vanguard visions may become sociotechnical imaginaries, but only after they gain widespread acceptance and achieve a measure of institutionalization.

Leaders of sociotechnical vanguards typically assume a visionary role, performing the identity of one who possesses superior knowledge of an emerging development and aspires to realize its more desirable potentials. Members of these vanguards typically share some commitments, such as to a future in which genomics plays a major role in biology, but they are rarely completely unified. They need not be scientists and engineers, and vanguards often include businesspeople, lawyers, activists, policy entrepreneurs inside government, and others. Whatever their background, they tend to fashion themselves as part of an avant-garde, riding and also driving a wave of change but competing with one another at the same time.[11] Because the boundaries of collective phenomena, such as social movements or scientific fields, are ambiguous, contested, and continually redefined in action, we cannot precisely delineate the vanguard's membership.[12] Formulating a vision and constituting a group to advance it is an imaginative act, not sharply separated from the activity of enrolling additional supporters, so vanguard visions are often reshaped as they gain support. Sociotechnical vanguards often include or split into factions that compete to advance visions that are only partially aligned. Yet even when their members pursue incompatible goals, sociotechnical vanguards can function as instances of what Maarten A. Hajer calls a "discourse coalition"—a collective that promotes partially shared meanings and incomplete storylines that, despite disunity, build an overall movement.[13]

The vanguard that envisioned the HGP took shape in the second half of the 1980s. In 1985 and 1986, a small group of elite scientists, most of them white men, initiated discussions of a project to sequence the human genome.[14] Initial visions called for a project narrowly focused on human genome sequencing. The proposal sought to revolutionize human genetics, which was engaged in the painstaking process of searching for genes one by one, by sequencing the whole genome.[15] The complete sequence would make it much easier to find genes and thus allow research to "shift from the question of how to find the genes to the question of what the genes do."[16] The vanguard behind this proposal had enough visibility in the US science policy community to provoke serious debate about whether the government should devote substantial funding to a genome project of some form, and it included scientists in the US Department of Energy (DOE), an agency with a large research budget that operates well-funded national laboratories. Extensive controversy took place about whether to launch a human genome project and, if so, of what form. Robert Cook-Deegan, a science policy analyst and participant in the debate, provides an informative account of how this initial vision was reformulated and won support in Washington.[17] Cook-Deegan shows how the vision of a program focused on human genome sequencing morphed into a proposal to proceed in stages, beginning with genome mapping and pilot projects to explore strategies for sequencing, but transitioning to full-scale sequencing only after major technological advances were achieved.

The new vision, most compellingly articulated in the National Research Council (NRC) report *Mapping and Sequencing the Human Genome* in 1988, also widened its scope: the project would analyze not only the human genome but also those of a number of "model" organisms, such as

Escherichia coli, yeast, the nematode, and the mouse.[18] (Table 2.1 shows the membership of the committee that wrote the report.) The institutional locus of the proposed project also shifted, moving from the DOE, the first agency to show serious interest, to a program jointly run by the National Institutes of Health (NIH) and the DOE. In 1988, the NIH director appointed James D. Watson, who was famous for the discovery of the double-helical structure of DNA and the head of the Cold Spring Harbor Laboratory, to direct the NIH program.[19] That year, the US Congress also agreed to fund the project, which was expected to cost $3 billion and take 15 years. The United States officially began its genome project in 1990 with the goal of completing it by 2005.

The vanguard supporting the HGP had important members in Europe and Japan. Although the proposal to sequence the human genome first emerged in the United States, the vanguard framed the HGP as an "international" initiative, even as it sold the project to the US Congress in part by

Table 2.1

NRC Committee on Mapping and Sequencing the Human Genome, 1988

Bruce M. Alberts (*chairman*)	University of California, San Francisco
David Botstein	Massachusetts Institute of Technology
Sydney Brenner	MRC Unit of Molecular Genetics, United Kingdom
Charles R. Cantor	Columbia University College of Physicians and Surgeons
Russell F. Dolittle	University of California, San Diego
Leroy Hood	California Institute of Technology
Victor A. McKusick	Johns Hopkins Hospital
Daniel Nathans	Johns Hopkins University School of Medicine
Maynard V. Olson	Washington University School of Medicine
Stuart Orkin	Harvard Medical School
Leon E. Rosenberg	Yale University School of Medicine
Francis H. Ruddle	Yale University
Shirley Tilghman	Princeton University
John Tooze	European Molecular Biology Organization, Federal Republic of Germany
James D. Watson	Cold Spring Harbor Laboratory

Source: NRC 1988, iii.

arguing that investing in genomics would enhance national competitiveness. As the US project was winning government funding at the end of the 1980s, genome programs of various types were also being launched in Europe. The European Community began a pilot project to sequence a yeast chromosome in 1989, with the idea of completing the whole yeast genome if the effort were successful. In France, the Association française contre les myopathies, a charitable organization, launched Généthon, a project focused on human genome mapping. The United Kingdom also established the Human Gene Mapping Program (HGMP) under the auspices of its Medical Research Council (MRC), although this program did not commit to sequencing the whole genome. To strengthen international coordination, the vanguard also founded an international professional society, the Human Genome Organization (HUGO) at the first "Genome Mapping and Sequencing" meeting at Cold Spring Harbor in April 1988.[20] The HGP thus had both international and national dimensions.

Nevertheless, the United States was the first and only government to make a commitment of billions of dollars to seeing the human genome mapped and sequenced. Ultimately, the International Human Genome Sequencing Consortium completed the human sequence.[21] However, the vast majority of the International Consortium's sequence data was produced by laboratories in only two countries, the United States and the United Kingdom, funded by the US NIH and DOE and the Wellcome Charitable Trust, a UK charity. China, France, Germany, and Japan also participated in the sequencing, and all of these countries, as well as others, made scientific contributions that advanced the broader genome project enterprise. To complicate matters further, a private company, Celera Genomics, was founded in 1998 for the announced purpose of launching a "private" effort to sequence the human genome before the "public" genome project could. This move provoked an acrimonious "race" between the two rival groups.

As a result of this complexity, one can speak of an "international" human genome project, of "public" and "private" genome projects, and of various national programs. To avoid confusion, I generally use the name "Human Genome Project" (and the acronym HGP) to refer to the overarching government- and charity-funded effort that began in the United States but grew into the International Consortium, while referring to nation-specific programs using more specific names (e.g., the UK Human Genome

Mapping Program, the US genome program, or the names of government agencies).[22] I refer to Celera's project using the company's name.

Building a New Biology

The advocates of the HGP aimed to make the human genome—and, more precisely, genomes in general—into tractable knowledge objects, building technology and data that would allow these objects to be analyzed and compared. The genomics vanguard sought to build the "infrastructure" needed to realize what Walter Gilbert at Harvard University called "a paradigm shift in biology."[23] This infrastructure would produce a wealth of data and techniques that would transform science. It would include technology for large-scale sequencing, vast quantities of genomic information, and algorithms and computing systems for analyzing genome data. Biology would become more mathematical and centered on the analysis of massive quantities of sequence data from many organisms and, in time, individuals.

Advocates of the HGP never saw completing the human sequence as a final endpoint; it was a step on the road to a new kind of biology. From the start, leading scientists predicted that as soon as the first sequence of the human genome was completed, sequencing capacity would be redirected to characterizing variation in the genome sequences of human populations and individuals—a prediction borne out by events. HGP leaders also told me early on that the sequencing of nonhuman organisms would not stop with the model organisms that the HGP had targeted. Agriculturally significant species would be sequenced, and one well-known scientist, anticipating what now goes by the name "geoengineering," speculated that blue-green algae and pine trees would be sequenced to help design solutions to global climate change.[24]

By the mid-1980s, when discussion of the HGP began, GenBank and the other government-funded sequence databases had become indispensable in molecular biology. Used in conjunction with a growing suite of computational tools, they had become much more than mere collections of data.[25] They had become a hybrid of data-storage and retrieval system, scientific instrument, and mode of scientific communication. In this role, they served to construct a wide range of "experimental systems" and "epistemic things."[26] They enabled scientists not only to distribute what Sabina Leonelli calls "small facts" but also to build common scientific ontologies,

Box 2.1

DNA and Genomes

The genomes of all living things consist of DNA, which scientists often represent as a linear molecule composed of two strands (figure 2.1). The two strands are weakly bonded together, so a double-stranded DNA molecule can separate into two single-stranded molecules without breaking. Each strand consists of a series of units lined up end to end, forming a chain or, more precisely, a double chain, with the two strands aligned such that each unit is "paired" with a corresponding unit on the other strand. The strands are composed of four different units, known as "nucleotides" or "bases," which scientists designate with the letters A, C, G, and T. These units may appear in any order, and the specific order found in a stretch of DNA is called its "sequence."

In keeping with the metaphors from cybernetics and information theory central to molecular biology, scientists liken a DNA molecule to a "text" written in a four-letter alphabet that conveys "information." In double-stranded DNA, A always pairs with T, and C always pairs with G. As a result, a single strand of DNA contains all of the information needed to create the strand that complements it. This redundancy is crucial to the replication of cells. When a cell divides to form two new cells, the strands separate, and each strand conveys all of the information encoded in the original DNA molecule to one of the new cells. The new cell then fills in the missing strand with correctly paired bases, and the process repeats itself as cells replicate. Scientists measure the length of DNA molecules (and quantities of genomic information) in bases (or base pairs for double-stranded DNA). The size of the genomes of the human and several model organisms appears in table 2.2.

The DNA of a genome contains sequences, known as "genes," that code for proteins as well as regulatory sequences that turn genes on and off (figure 2.2). Genes are made up of "exons"—the parts that are translated into protein—and may also contain introns, which are removed before the protein is assembled. Genomes also contain regions of unknown function, which many biologists classified as "junk" at the outset of the HGP. In 1988, the human genome was believed to contain on the order of 100,000 genes. The number of genes is now believed to be about 21,000.

Scientists have incorporated the properties of DNA into the technologies for investigating it. For example, base pairing allows scientists to detect specifically targeted sequences using a technique called "hybridization." Single-stranded DNA can be labeled with a radioactive tag, creating a "probe." If the radioactive probe encounters its complementary sequence, it will bind (or "hybridize") to it, making it detectable using X-ray film (figure 2.3). Scientists thus can test a DNA sample for the presence or absence of a specific sequence.

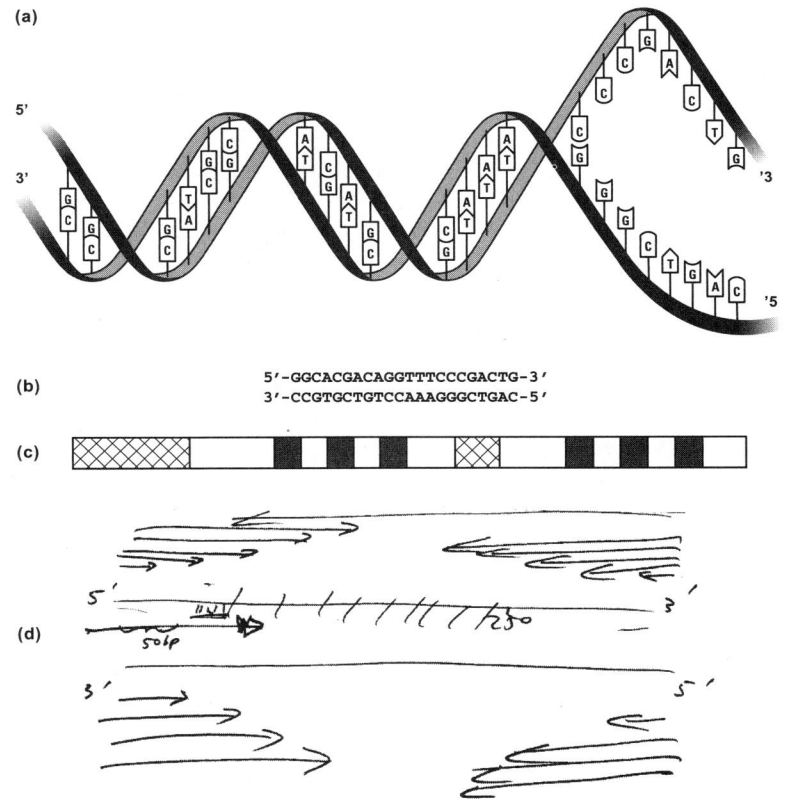

Figure 2.1
Four ways genome scientists represent DNA. (a) Double-helical structure showing base pairing and a specific sequence. The strands of the DNA molecule can separate without changing the order of the bases, as shown on the right. **(b)** A double-stranded DNA sequence displayed as a string of text. The human genome is 140,000,000 times longer than this 22-base-pair sequence. If printed in a linear array (12 letters per inch), the human genome sequence would easily reach from Cambridge, Massachusetts, to Cambridge, England—the cities where practicable DNA-sequencing methods were developed in the 1970s. The sequence shown is from the paper that first described Maxam-Gilbert sequencing in 1977 (Maxam and Gilbert 1977). **(c)** A representation of a genetic sequence with different "regions" or "features" of the sequence indicated using shading. **(d)** Strands of DNA depicted as lines in a sketch from a laboratory discussion in 1991.
Illustration by Chris Cooley, Cooley Creative, LLC.

A recombinant DNA technique known as cloning has been a key method in biotechnology since the early 1970s. Cloning allows scientists to make huge numbers of copies of interesting DNA fragments and thus to acquire sufficient material to study. The process involves inserting a DNA fragment of interest (say, a piece of mouse or human DNA) into a "cloning vector" that transfers it into the cell of a microorganism easily cultured in the laboratory (figure 2.4). The microorganism containing the "insert" is called a "clone." Once a fragment of human DNA has been successfully cloned, researchers can produce many copies of it by growing the clone—which is alive—using tissue-culture techniques. When the organism replicates itself, the inserted DNA will also be reproduced. Working with clones is a meticulous business because each clone must be separately stored and protected from contamination. However, because clones can be preserved in freezers, shipped among laboratories, and copied by growing the organism, they constitute DNA fragments as standardized and reproducible knowledge objects for use in ongoing research. Clones became an indispensable tool in molecular biology, and so that scientists could build on the work of others, a norm developed requiring authors to make clones used in published papers available to qualified scientists on request.

Table 2.2
Sizes of Genomes

Organism	Estimated Genome Sizes (base pairs)	Number of Genome Equivalents Sequenced (December 1991)
Escherichia coli (bacterium)	4,600,000	0.597
Saccharomyces cerevisiae (yeast)	12,000,000	0.203
Caenorhabditis elegans (nematode or "worm")	100,000,000	0.007
Drosophila (fruit fly)	120,000,000	0.018
Mus musculus (mouse)	3,000,000,000	0.002
Homo sapiens (human)	3,000,000,000	0.005

Note: Estimated (haploid) genome sizes toward the beginning of the HGP. Only a small fraction of the genomes of most of these organisms had been sequenced as of December 1991.
Source: Cinkosky, Fickett, Gilna, et al. 1992, 272.

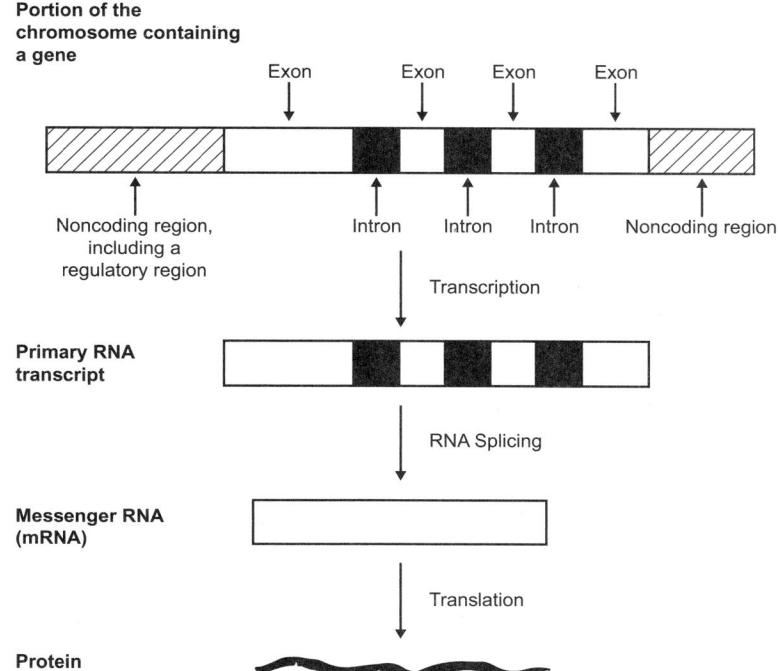

Figure 2.2
How genes are expressed in human cells. This figure and caption appeared in a key NRC report *Mapping and Sequencing the Human Genome* (1988): "How genes are expressed in human cells. Each gene can specify the synthesis of a particular protein. Whether a gene is off or on depends on signals that act on the regulatory region of the gene. When the gene is on, the entire gene is transcribed into a large RNA molecule (primary RNA transcript). This RNA molecule carries the same genetic information as the region of DNA from which it is transcribed because its sequence of nucleotides is determined by complementary nucleotide pairing to the DNA during RNA synthesis. The RNA quickly undergoes a reaction called RNA splicing that removes all of its intron sequences and joins together its coding sequences (its exons). This produces a messenger RNA (mRNA) molecule. The RNA chain is then used to direct the sequence of a protein (translation) according to the genetic code in which every three nucleotides (a codon) specifies one subunit (an amino acid) in the protein chain." *Source:* NRC 1988, fig. 2-3, p. 18. Reprinted with permission by the National Academy of Sciences. Courtesy of the National Academies Press, Washington, DC.

(a)

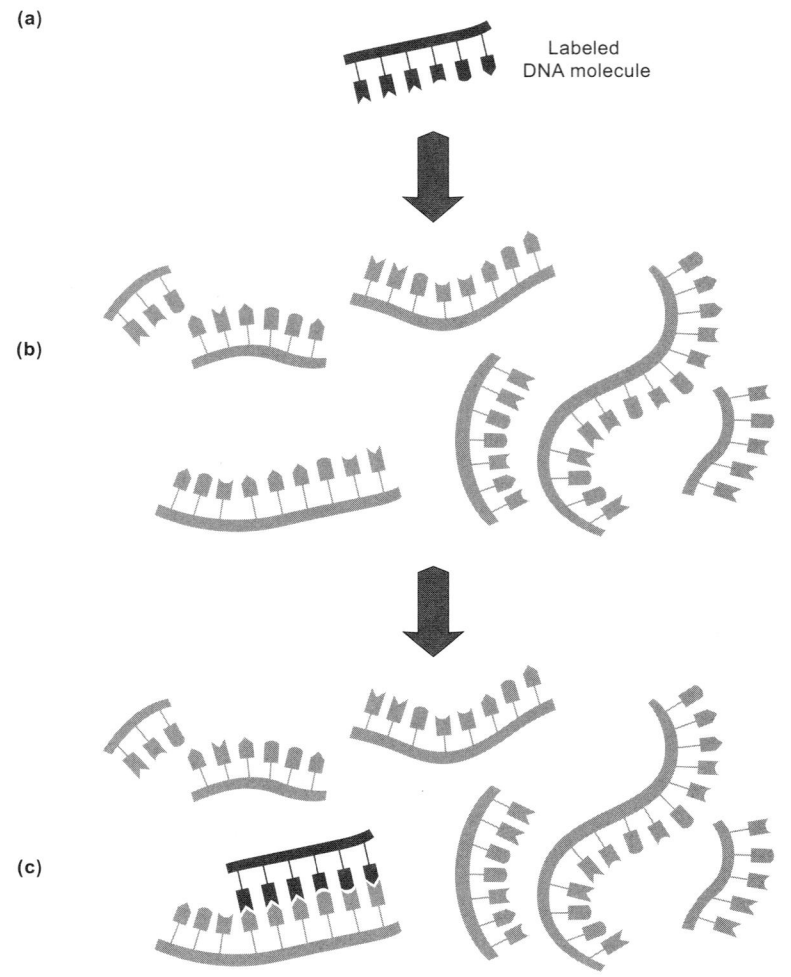

(b)

(c)

Figure 2.3
Detecting a specific DNA sequence with a labeled probe. Scientists can exploit the base-pairing property of DNA to test samples for the presence of specific sequences. A radioactively labeled "probe" containing the sequence of interest is prepared **(a)**. The probe is poured over the DNA to be tested **(b)**. The probe binds only to sequences that match the probe **(c)**. This technique for detecting specific sequences is called "hybridization." Fluorescent tags can also be used. Illustration by Chris Cooley, Cooley Creative, LLC.

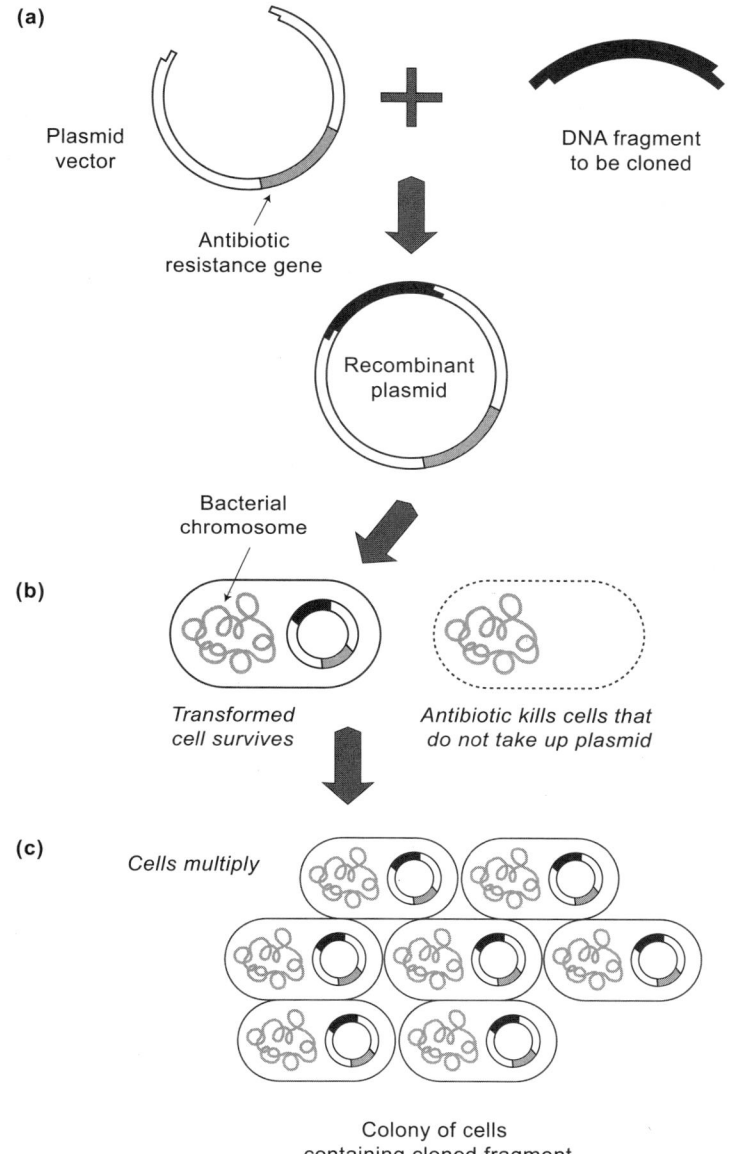

(a)

Plasmid vector

DNA fragment to be cloned

Antibiotic resistance gene

Recombinant plasmid

Bacterial chromosome

(b)

Transformed cell survives

Antibiotic kills cells that do not take up plasmid

(c)

Cells multiply

Colony of cells containing cloned fragment

Figure 2.4
Cloning DNA using recombinant DNA technology. A small, circular piece of DNA called a plasmid is used as a "vector" to transfer a DNA fragment into a bacterial cell. The plasmid, which contains a gene conveying antibiotic resistance, is cut, and the DNA fragment to be cloned is spliced in, as shown in **(a)**. Some bacterial cells take up the plasmid. After exposure to the antibiotic, only cells containing the plasmid survive **(b)**. The cells multiply, producing many copies of the cloned fragment **(c)**. The cells can be grown, stored, and transported, providing an ongoing supply of the cloned DNA. Illustration by Chris Cooley, Cooley Creative, LLC.

analytic practices, and modes of communication that linked researchers studying a wide variety of organisms and biological functions.[27] Hallam Stevens argues that the study of sequences constituted nothing less than a new vision of life.[28]

After sequence data started to accumulate in the late 1970s and 1980s, scientists began to ask new sequence-based questions. How, for example, did the DNA coding for human insulin differ from bovine insulin or mouse insulin? They also used sequence databases to examine change over evolutionary time, developing new, DNA-based ways to redraw the "tree of life." Algorithms for searching the databases allowed scientists to see what they could learn about newly found sequences. Comparing a new sequence from, say, yeast with all the sequences in the databases might indicate that it was, for example, the yeast's version of a gene also found in *E. coli*. In this manner, searches for "homologous" sequences could yield clues about gene function. The more complete the database, the more useful because the more information it contained about a variety of organisms, the more likely that searches of the database would yield valuable insights. The HGP thus would not only identify all of the individual human genes but also enable computational analysis that would transform every subfield of the biosciences.

Mapping and Sequencing the Human Genome also imagined making the data freely accessible to the entire scientific community. The plan was to house the sequences of the human and a number of other organisms in the government-funded biological databases that had been established at the beginning of the 1980s. The fact that the utility of a sequence database increases with its size and completeness had encouraged highly centralized organization; in effect, the ultimate database would be a single, comprehensive one containing a collection of all of the sequences (from many organisms) that the global scientific community had produced. Largely for nationalistic and geopolitical reasons, however, governments had established three databases on different continents—namely, GenBank in the United States, the European Molecular Biology Laboratory's Nucleotide Sequence Data Library, and the DNA Data Bank of Japan. These three databases cooperated extensively, with each sending its data to the others, an arrangement that produced three essentially equivalent collections. In keeping with the vision of an "international" project, most of the vanguard expected the HGP to greatly expand the coverage of all three databases. In contrast, Gilbert envisioned sequence databases as the basis for a commercial business. He imagined copyrighting DNA sequences and selling access to them, but his effort to set up a genome-sequencing company in the late 1980s failed to attract capital.[29]

The HGP vanguard drew internal inspiration and outside support from widely shared sociotechnical imaginaries of desirable futures. Its vision drew on several well-established sociotechnical imaginaries, including the imaginary of science as quest for knowledge and an exploration of exciting new frontiers. The vanguard also drew on the imaginary of investment in science as a source of economic growth and national competitiveness. It tapped into the imaginary of biomedical science bringing cures and conquering dreaded disease. The promise of realizing these aspirations motivated genome scientists as individuals and helped win institutional support from the Congress, the research community, and the biomedical, biotechnology, and pharmaceutical industries.[30] The genomics vanguard packaged the HGP to appeal to various constituencies. Its very name, "Human Genome Project," was a streamlined slogan, not a literal description. Emphasizing the *human* genome (rather than genomes in general) stressed medical applications and cast the project as a quest for self-understanding. Framing the goal of sequencing "the" human genome (as if such a unitary object existed) suggested finality and closure. Calling the enterprise a "project" (as if it had a clear beginning and end) also suggested a finite task that would be "completed," once and for all.

As the proposal for an HGP gained policy traction, members of the vanguard coined the term *genomics*, using it to capture their vision of a new science that would move beyond traditional molecular biology by making whole genomes into tractable scientific objects. At the end of 1987, the inaugural issue of the scientific journal *Genomics* appeared. An editorial titled "A New Discipline, a New Name, a New Journal" announced the arrival of genomics and outlined an early vision of the field. By way of explaining the choice of the neologism *genomics*, the editors wrote that the suffix *-ology* "suggests academic isolation," whereas the more energetic *–ics* "suggests a method of attack on life's problems."[31] Genomics was soon associated not only with an emerging research domain but also with the idea that the move "from genetics to genomics" was a thrilling paradigm shift that demanded investments of money, talent, and hope.[32]

The discourse of hope was especially prominent in discussion of bio-medical applications. Groups representing families suffering from inherited genetic disorders, such as Huntington's disease, were already actively supporting research to find disease-causing genes, gathering DNA from affected families, and even funding and directing research.[33] For such groups, genomics promised to find disease genes and speed the investigation of biological mechanisms and, ultimately, development of new and more effective drugs. Genome project leaders expected that the HGP's greatest contributions to improving health would come from knowledge about common conditions—such as cancer, heart disease, hypertension, diabetes, and asthma. Genome research would yield a deeper understanding of both normal and pathological human biology, leading to genetic tests for predispositions to common diseases and ultimately to effective prevention or cures.

Sociotechnical Challenges

The goal of the genomics vanguard—to create knowledge and technology that would revolutionize biology and medicine in the twenty-first century—was in no way a modest one. At the close of the 1980s, as science policy makers grew increasingly supportive of funding a concerted effort to analyze the human genome, even the HGP's most enthusiastic advocates acknowledged that the project would have to overcome major obstacles. Existing genome technology was uniformly perceived as inadequate, so the project was predicated on catalyzing technological change. When the project began, the only genomes sequenced completely were those of viruses—tiny in comparison to the human genome. Indeed, the first sequencing of the genome of a free-standing organism was not completed until 1995, when a team lead by J. Craig Venter published the sequence of *Haemophilus influenzae*, a bacteria with a genome about $\frac{1}{1,600}$ the size of the human genome.[34] Mapping and sequencing techniques were labor intensive and time consuming, and many scientists feared that the HGP would become a financial boondoggle, contending privately and sometimes publicly that sequencing the human genome would waste precious resources.[35] To many observers, neither extant technology nor extant modes of governance seemed capable of realizing the vanguard's vision.

In the late 1980s, the only workable sequencing technologies were based on gel electrophoresis (figure 2.5). Determining the 3 billion base-pair sequence of the human genome would entail sequencing all of the 24 different human chromosomes (chromosomes 1 to 22 and the X and the Y).[36] Each chromosome, a single DNA molecule ranging in size from about 50 million to 250 million base pairs, dwarfed the total volume of human sequence information (some 2 million base pairs) that had accumulated in the centralized databases, such as GenBank.[37] Electrophoresis could only sequence short strips of DNA—known as "reads"—a few hundred base pairs in length. By creating overlapping DNA fragments from a specific region of a genome and subsequently sequencing these fragments, scientists could join short reads into longer, contiguous sequences (figure 2.6). Using this approach, genome researchers had sequenced a number of human genes, but attacking the entire genome with these techniques would require tens of millions of reads. Moreover, to produce just a handful of reads, skilled laboratory workers had to meticulously perform many time-consuming steps. Not surprisingly, the NRC Committee on Mapping and Sequencing the Human Genome concluded that "the cost and inefficiency of current DNA sequencing technologies are too great to make it feasible to contemplate determining the 3 billion nucleotides of the DNA sequence of the human genome within a reasonable time. The largest contiguous segment of human DNA determined to date is the 150,000 nucleotides encoding the human growth-hormone gene. This is 0.005 percent of the total genome."[38]

In light of this gap between goals and capabilities, the NRC report recommended governing the HGP by proceeding in phases. The project should initially focus on constructing basic maps of the genome, which the NRC committee expected to become immediately useful to a wide range of biomedical research. Pilot sequencing projects would also be funded, mainly with the hope of gaining experience, developing new techniques, and, in time, making substantial progress on some of the model organisms. A transition to large-scale sequencing would occur only in the wake of substantial improvements in technology.

Genome mapping also posed major challenges. Like geographic maps, genome maps can be constructed at different levels of detail. At one extreme stand stylized depictions of chromosomes that assign numbers to the light and dark bands visible under a microscope. At the other extreme

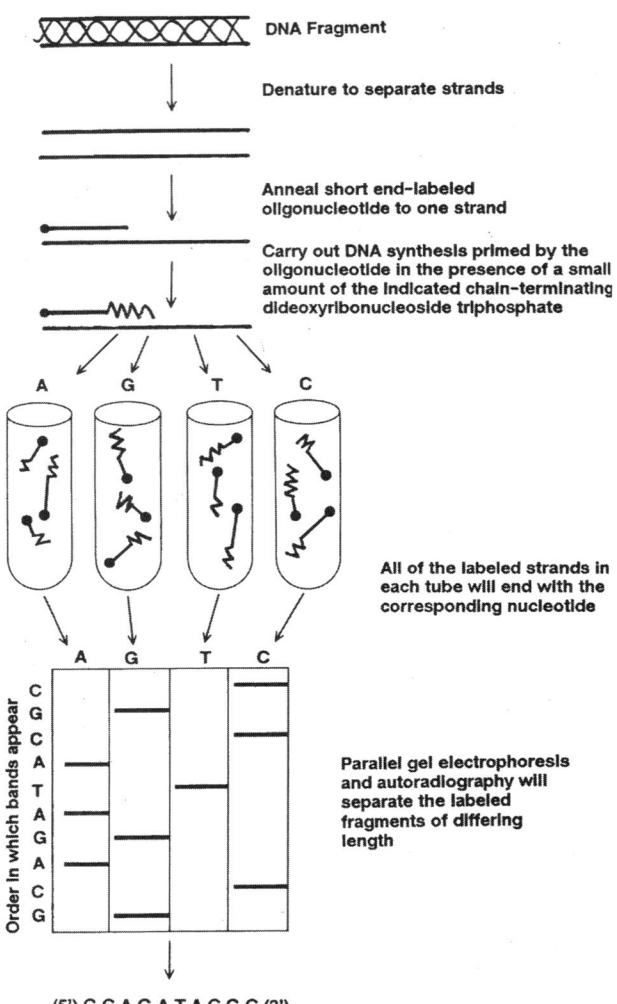

DNA Fragment

Denature to separate strands

Anneal short end-labeled oligonucleotide to one strand

Carry out DNA synthesis primed by the oligonucleotide in the presence of a small amount of the indicated chain-terminating dideoxyribonucleoside triphosphate

All of the labeled strands in each tube will end with the corresponding nucleotide

Parallel gel electrophoresis and autoradiography will separate the labeled fragments of differing length

Order in which bands appear

(5') G C A G A T A C G C (3')

Sequence of end-labeled strand

is the complete sequence—the "map at the highest level of resolution"—that the scientists in the late 1980s hoped someday to produce.[39] In between, there are several kinds of lower-resolution maps, each of which measures different things (figure 2.7). These maps include "genetic linkage maps" ("genetic maps," for short), which are based on patterns of inheritance in multigenerational families.[40] To build these maps, scientists collect data on the transmission of "polymorphic markers" from parents to offspring. Polymorphic markers are simply short sequences that come in several forms, making them traceable across generations. Two markers that are likely to be inherited together are "linked," and the probability that they will be co-inherited becomes a measure of "genetic distance." Determining the position of a gene on a genetic linkage map is an important step in finding a disease gene because it localizes the gene to a particular region of a specific chromosome. A very coarse linkage map of the human genome was published in 1987, and the NRC committee was confident that a concerted mapping program could produce a much more detailed map within several years.[41]

Linkage maps display the order of a set of polymorphic markers along a chromosome but cannot measure the physical distance separating them. Accordingly, researchers also were building several kinds of "physical

---

Figure 2.5

Sanger sequencing. This figure, which describes DNA sequencing by the "enzymatic method," appeared in the NRC report *Mapping and Sequencing the Human Genome* (1988). This method takes advantage of an enzyme that participates in the replication of DNA in living cells. The enzyme (known as DNA polymerase) uses a piece of single-stranded DNA as a template and creates a complementary strand, building up a chain of bases in the correct order. Each new base is attached one by one to the end of the growing chain. Molecular biologists use DNA polymerases to make complementary strands of DNA in vitro. In Sanger sequencing, a polymerase is used to build complementary strands that are artificially interrupted by including an altered nucleotide (dideoxynucleoside triphosphate) that blocks the addition of the next nucleotide in the chain. The result is a family of DNA fragments that vary in length by one nucleotide. These fragments are sorted by length using gel electrophoresis, which makes it possible to read the sequence by inspecting the gel. A single "run" of such a sequencing gel produces a strip of sequence—or "read"—that is only about 500 bases in length. Longer sequences can be assembled by combining reads (figure 2.6). *Source:* NRC 1988, fig. 5-1, p. 63. Reprinted with permission by the National Academy of Sciences. Courtesy of the National Academies Press, Washington, DC.

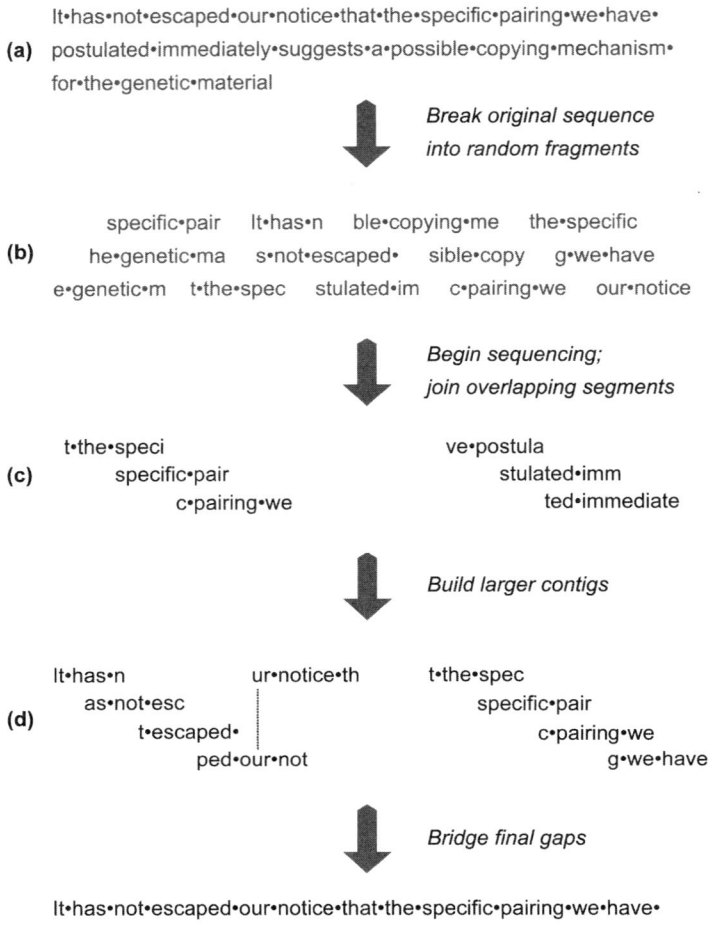

(a) It·has·not·escaped·our·notice·that·the·specific·pairing·we·have· postulated·immediately·suggests·a·possible·copying·mechanism· for·the·genetic·material

Break original sequence into random fragments

(b) specific·pair   It·has·n   ble·copying·me   the·specific
he·genetic·ma   s·not·escaped·   sible·copy   g·we·have
e·genetic·m   t·the·spec   stulated·im   c·pairing·we   our·notice

Begin sequencing; join overlapping segments

(c) t·the·speci   ve·postula
specific·pair   stulated·imm
c·pairing·we   ted·immediate

Build larger contigs

(d) It·has·n   ur·notice·th   t·the·spec
as·not·esc   specific·pair
t·escaped·   c·pairing·we
ped·our·not   g·we·have

Bridge final gaps

(e) It·has·not·escaped·our·notice·that·the·specific·pairing·we·have· postulated·immediately·suggests·a·possible·copying·mechanism· for·the·genetic·material

Figure 2.6
Connecting overlapping reads to form a continuous sequence. Short "reads" of DNA sequence can be aligned and joined to form larger strips of sequence. The principle is illustrated here using a sequence of English-language characters rather than the four-letter "alphabet" of DNA. (This example uses a "read length" of about 12 characters rather than about 500 in DNA sequencing.) First, many copies of an unknown sequence (**a**), shown in gray to indicate that the sequence is unknown, are broken at random places to form small fragments (**b**). Sequencing of these fragments then begins, and as each new "read" is produced, its sequence is compared with the

maps." Physical mapping entailed working with large collections of cloned DNA fragments, especially cosmid clones and yeast artificial chromosomes (YACs). Scientists refer to a large collection of clones derived from the same DNA source, such as the human genome, as a "library." Unlike the orderly collections of books that the term *library* brings to mind, however, the libraries of clones available at the outset of the HGP were essentially uncataloged; each clone would be stored in its own container, but the containers were in random order. Bringing order to these collections of random fragments was a major goal of the mapping stage of the HGP, and the task was expected to take years. To "cover" the human genome—that is, to include DNA fragments representing the entire human genome—would require thousands to hundreds of thousands of individual clones (depending on the kind of clone and the desired level of redundancy).

One type of physical map is called a "restriction map" (or "macrorestriction map" to emphasize scale).[42] The largest DNA molecule mapped using this technique when the NRC committee completed its deliberations was that of *E. coli*, $\frac{1}{650}$ the size of the human genome.[43] Another kind of physical map—a "contig map"—is based on breaking a genome into many partially overlapping fragments, cloning those fragments, and assembling a contiguous set of overlapping clones that reflects their original order on the chromosome (figure 2.8). Contig maps of such organisms as *E. coli* and the nematode *Caenorhabditis elegans* were under construction when the NRC wrote its report.[44] Constructing contig maps of human chromosomes was believed to be possible, although not without considerable investments of funds and effort.

---

reads already completed. Overlapping reads are joined, creating small contigs (c), shown in black to indicate that their sequences are now visible. As sequencing continues, new reads extend existing contigs (d). Final gaps are eventually filled, revealing the originally unknown sequence in its entirely (e). The DNA sequencing technology at the outset of the HGP produced reads of some 200 to 500 base pairs, typically containing some errors, especially toward the end of the read. The human genome contains many repetitive sequences, which complicate the problem of assembling continuous sequence in the correct order. The sentence shown is a famous line from Watson and Crick's paper from 1953 describing the structure of DNA. Illustration by Ranjit Singh.

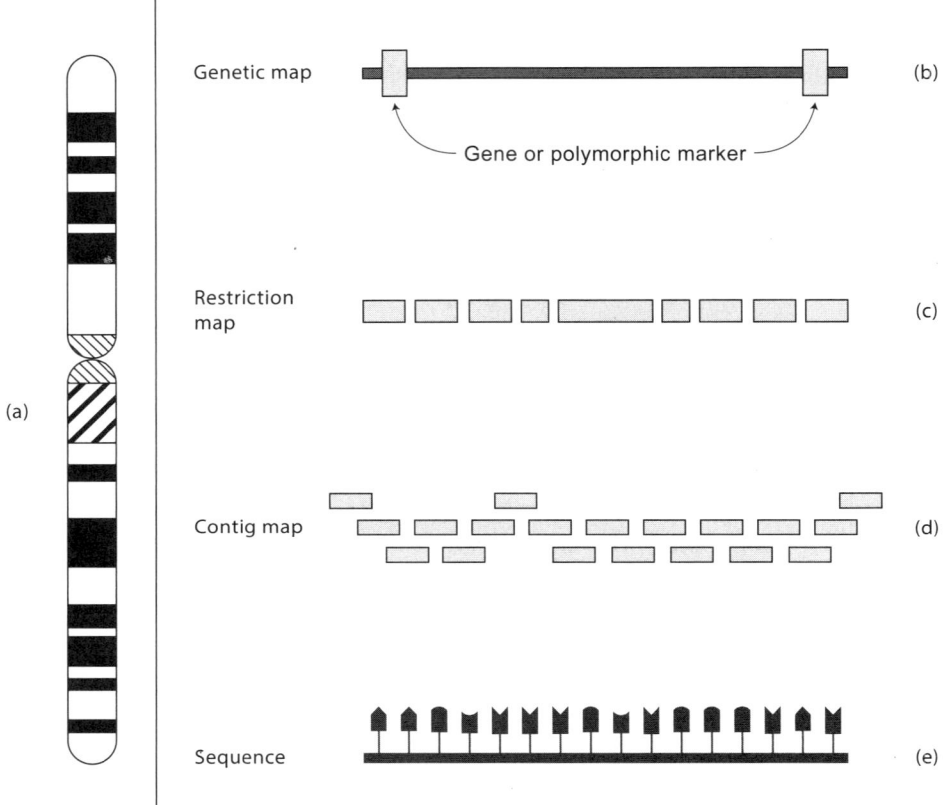

Figure 2.7
Maps at different levels of resolution. The lowest-resolution genome maps depict banding patterns on chromosomes, such as the map of human chromosome 16 shown in **(a)**. As the HGP took shape, the genomics vanguard imagined building maps at very different levels of resolution, with genome sequences **(e)** conceptualized as maps at the highest level of resolution: one base pair. They also envisioned building contig maps **(d)**, seeking to span whole chromosomes with overlapping clones, the resolution depending on the cloning technology used: for cosmids, roughly 40,000 base pairs; for YACs, from 100,000 to 1,000,000 base pairs or more. Genome scientists also planned to build restriction maps of whole chromosomes **(c)**, with an expected resolution of 1 to 2 million base pairs. The resolution of genetic linkage maps cannot be expressed in terms of base pairs because linkage maps and physical maps measure fundamentally different things. Distances in linkage maps are measured in centimorgans, which indicate the degree of linkage between polymorphic markers or genes **(b)**. On average, one centimorgan corresponds to a physical distance of about 1 million base pairs. Illustration by Chris Cooley, Cooley Creative, LLC.

The NRC expected that physical mapping would create an ordered collection of clones to provide a starting point for sequencing. Producing such an ordered collection would require working with tens or hundreds of thousands of clones. In this context, genome researchers were preoccupied with finding ways to analyze more DNA more quickly, less expensively, more easily; to simplify the "jigsaw puzzle" problem of assembling contigs by manipulating bigger DNA fragments; and to extract more information from the same amount of work.[45] Increasing "throughput"—the speed of production—was a central goal. Michael Fortun does not exaggerate when he writes that an obsession with speed is constitutive of genomics.[46] Indeed, the NRC report presented ongoing increases in efficiency as among the distinguishing characteristics of the HGP: "The human genome project should differ from present ongoing research inasmuch as the component subprojects should have the potential to improve by 5- to 10-fold increments the scale or efficiency of mapping, sequencing, analyzing, or interpreting the information in the human genome."[47]

The project's supporters expected to achieve such increases through some combination of incremental improvements in existing techniques, automation, organizational scale-up, and the creation of new informatics systems for establishing and maintaining control over vast quantities of genomic data.[48] In the early years of the HGP, many members of the vanguard also expected major technological breakthroughs to be necessary. They doubted that what was then the only practical method for sequencing DNA—gel electrophoresis—would ever be sufficiently fast and inexpensive to sequence large genomes, and they expected that large-scale sequencing would require entirely new technologies based on scanning tunneling microscopy, single-molecule detection, the use of hybridization arrays, or some sort of DNA microchip. A few members of the vanguard argued, however, that electrophoresis could do the job and that the delays inherent to developing radically new technologies meant that such "exotic," "pie in the sky" approaches remained far in the future.[49]

Organizational Complexities

The genomics vanguard expected the HGP to require not only technological advances but also change in scientific practices and institutions; some influential members of the HGP leadership argued in the early 1990s that some of the most important impediments to the project were institutional

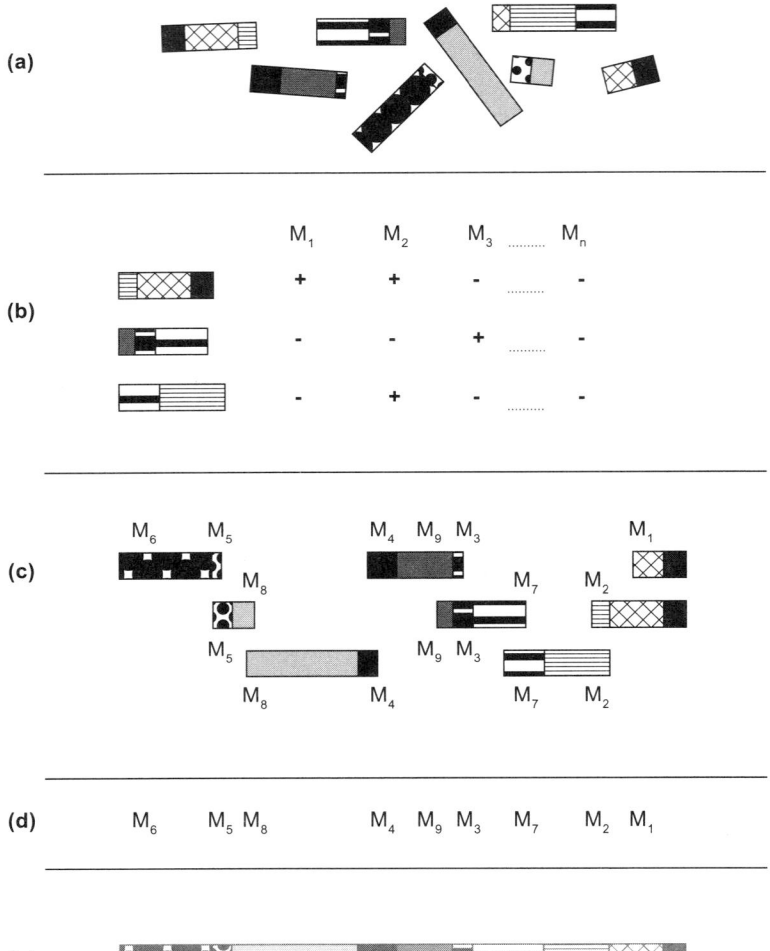

Figure 2.8
Assembling a contig map by identifying overlapping clones. At the outset of the HGP, scientists worked to produce contig maps using several methods to find over-lapping clones. This simplified schematic illustrates one method. Many copies of a genome are cut randomly into large fragments (tens or hundreds of thousands of base pairs) **(a)**. These fragments are cloned, and a library of clones is created, some of which overlap. Each clone is systematically tested against a set of markers (M_1 to M_n), such as radioactively labeled DNA "probes," to find overlaps. A "positive" test result means that the clone contains the marker. Like the order of the clones, the order of the markers on the chromosome is initially not known. However, test results **(b)** re-veal that some clones share a particular marker, indicating that they overlap. In this

ones. The challenge of building and governing organizations capable of mapping and sequencing on a large scale was a critical dimension of the problem. Like the rest of molecular biology, human genetics and pre-HGP genetic mapping was a world of small, "independent" laboratories. The HGP vanguard considered these "cottage industry" laboratories incapable of high-throughput mapping and sequencing, and they imagined building "genome centers" to take advantage of economies of scale.

The challenges were complicated by intense competition. Human genetics, a domain of research partially overlapping the emerging field of genomics, was especially competitive. In the late 1980s and early 1990s, human geneticists focused on finding the genes underlying the relatively rare diseases passed down through simple Mendelian inheritance. Even the most visible of these "single-gene disorders," such as cystic fibrosis, Huntington's disease, neurofibromatosis, Tay-Sachs disease, and various forms of muscular dystrophy, were rare in comparison to the common diseases of greatest importance to public health. However, finding such genes was possible with the techniques available in the 1980s, although difficult and uncertain to succeed. Genome scientists compared gene hunting to looking for a needle in a haystack. Fiercely competitive, zero-sum races developed in many cases because a relatively small number of Mendelian disorders provided attractive gene-hunting targets. Prestige, grants, patents, and years of work were all at stake in these races. Human geneticists therefore tightly controlled access to biomaterials, DNA "probes" used to hunt for genes, and other data, providing them to other laboratories only cautiously and often after considerable delay.

---

example, the first and third clones overlap because they both test positive for marker M_2. As testing proceeds, the alignment of overlapping clones creates small contigs that join to form larger contigs as additional test results accumulate. In theory, each chromosome will eventually be spanned by a single contig because a chromosome is a single continuous DNA molecule. A contig is depicted in (c). The order on the chromosome of both the clones (c) and the markers (d) is thus established, although the exact location of neither is determined. The map shown in gray in (e), which displays the contig with overlaps eliminated, cannot be completed using this technique because the extent to which each clone overlaps its neighbors cannot be ascertained. However, the ordered clones (c) and ordered markers (d) can serve as starting points for more detailed mapping or sequencing. Illustration by Chris Cooley, Cooley Creative, LLC.

Against this backdrop, the genomics vanguard stressed that the HGP would require unprecedented "sharing" of data and materials among laboratories. As the NRC report put it, "The human genome project will differ from traditional biological research in its greater requirement for sharing materials among laboratories. For example, many laboratories will contribute DNA clones to an ordered DNA clone collection. These clones must be centrally indexed. Free access to the collected clones will be necessary for other laboratories engaged in mapping efforts and will help to prevent a needless duplication of effort."[50] Instituting this form of cooperation would require the fledgling HGP to effect significant change in knowledge-control regimes.

Scientists critical of the project also posed a problem. In addition to worrying about wasting funds, some biologists contended that the HGP would tie up talent in uninteresting and repetitive work because mapping and sequencing procedures would have to be performed on huge numbers of samples. They also worried that "megaprojects are inviting targets for the political control of science."[51] The successes of molecular biology, they argued, had stemmed from "independent" scientists pursuing their personal hypotheses without bureaucratic coordination. A "centralized" genome project could damage this culture and harm the careers of young scientists.

The genomics vanguard responded that funds would be carefully managed and that the HGP would ultimately save money. They pointed out that the expected cost of the HGP—$3 billion distributed over 15 years, or about $200 million per year—was small in comparison to the total NIH budget. They also argued that genome data and technology would *reduce* the amount of boring work that biological research would require. Automation would eliminate drudgery. Genome maps and sequence data would give every new investigation a head start. Systematic mapping would speed the hunt for genes, solving the needle-in-the-haystack problem by cataloging all of the pieces of straw. Such arguments proved persuasive enough to win the funds required to launch the project, but skeptics held their ground. Finding ways to build regimes capable of keeping the HGP on course—of regulating expenditures, making demonstrable progress, and maintaining accountability—became a prominent concern among the HGP leadership.

The HGP also attracted critics with ethical and political concerns, and some genome scientists complained that the project had become a "lightning rod" for broader worries about biotechnology. Observers expected that genome research would yield predictive tests for genetic diseases long before effective therapies would become available. Patients, the NRC committee wrote, would have to decide whether it is better "to know one's fate when it is out of one's control."[52] Questions about "genetic privacy" and "genetic discrimination" received considerable attention in policy discussions, and the prospect that genetic test results would lead to denials of life insurance—and health insurance in the United States—grew salient.[53] The potential for a "new eugenics" also sparked concern, as did reservations about the possibility that genetic research might encourage an excessively deterministic view of human nature or reinforce racist and sexist ideologies.[54] Like the leaders of explicitly political revolutions, the genome vanguard thus faced the question of how to assure potential supporters that its revolution would not lead to undesirable outcomes.

James D. Watson advanced one answer to this question at the press conference announcing his appointment as director of the NIH genome program in 1988. Watson called for building research on the societal issues raised by genomics into the HGP research program, and he publicly committed the NIH to devoting 3 percent of its budget to "the ethical and social implications area." Beyond ensuring that "society learns to use the information only in beneficial ways," Watson warned that a failure to act could lead to "abuses" and provoke "a strong popular backlash against the human genetics community."[55] The result was the US Ethical, Legal, and Social Implications (ELSI) program, which aimed to "anticipate and address" emerging societal issues and provide policy guidance.[56] Parallel programs were established in Europe (known as ELSA, with the A standing for "Aspects"). The 3 percent budget, later raised to more like 5 percent, made ELSI the best-funded bioethics program in US history.

Instituting ELSI conveyed the promise that the genomics vanguard would not neglect the ultimate societal effects of genomics, and the move was well received in the press. The ELSI vision expressed the hope of escaping the so-called Collingridge dilemma. David Collingridge argued that an intractable double bind stymies efforts to exert societal control over emerging technology: in the early stages of development of a technology, he argued, too little is known to predict its social effects; later, when the technology has achieved widespread use, its societal impact may be known, but the commitment is irreversible.[57] In effect, the ELSI concept advanced a new vision for integrating well-funded research about ethical, legal, and social implications into the early phases of major scientific projects.

Not surprisingly, the question of precisely what the ELSI program should attempt to accomplish immediately became a matter of controversy.[58] Some observers saw ELSI as reflecting "alarmist hype" about research that raised no truly novel issues, and high-ranking NIH officials questioned the wisdom of spending large sums to subsidize "the vacuous *pronunciamentos* of self-styled 'ethicists.'"[59] Other observers saw ELSI as a political strategy to mitigate opposition to genome research—a view that Watson's warning about a popular backlash seemed to support. Bioethicists, legal scholars, historians of science, and others predicted that ELSI research calling for cautious consideration of the downsides of the HGP would be labeled "uninformed" and argued that the premise that the HGP was a prima facie good would unduly constrain the scope and depth of inquiry.[60]

Despite these debates, the commitment of significant funding to research on ELSI issues was widely seen as demonstration of how seriously the genomics vanguard took matters of ethics, and many observers believed Watson's declaration helped win congressional support for the fledgling genome program. More deeply, ELSI's emphasis on "implications" and "misuse" framed political and societal issues as embedded not in the fundamental biological research that the vanguard proposed but in the *application* of genomic knowledge. This view of ELSI reinforced the modernist imaginary of science as a space properly separated from society and politics. The program thus provided a visible public demonstration of the genomics vanguard's commitment to addressing the societal implications of the HGP, while also framing those implications as pertaining more to unrealized applications than to the goals, social organization, and control of the scientific research itself.[61] It thus fostered a vision of a revolution that would transform the world yet do so in beneficial and relatively orderly ways.

Changing Laboratories

Mapping Laboratory, 1988

To get a sense of how knowledge was made in a genomics laboratory at the outset of the HGP, let us consider one laboratory that was part of the vanguard of the proposed project. In 1988, MAPPING LABORATORY, which was housed in a US research university, looked much like the other laboratories in the department: the same physical plant, the familiar laboratory benches, the same types of mostly small, bench-top equipment. It occupied a bit more space, divided into several laboratory and office rooms, but one could not locate the boundary separating Mapping Lab from the other laboratories on the floor from its appearance alone. Mapping Lab had no robots or DNA-sequencing machines. The offices had personal computers for word processing and making diagrams, but no computers were found in the wet-laboratory rooms. The one exception was a computer terminal used occasionally to search GenBank. The fax machine was busy; email had not yet caught on among biologists, although a few laboratory members did know how to use it.

Like other genomics laboratories operating at this time, Mapping Lab conformed relatively closely to the patterns that Karin Knorr Cetina describes in her study of the "epistemic culture" of molecular biology.[62] As she shows, molecular biology has a highly individuated culture, and research takes place in small laboratories, typically consisting of a laboratory head and few postdoctoral researchers, graduate students, and technicians. The knowledge-control regime governing these laboratories vests epistemic and political authority in the "laboratory head"—figured as an autonomous individual who personifies the laboratory, speaks for it, and directs its research. People work mostly alone at their individual benches. Benchwork, a skilled craft, requires careful manipulation of small tools and fragile materials. Protocols for processing materials and performing assays resemble "complicated cooking" with many steps.[63] The work necessarily has a collective dimension, and papers with multiple authors are the norm. Research is divided into "projects" generally conducted by a postdoctoral scientist or graduate student (sometimes with assistance from other junior researchers or a technician or two) under the supervision of the laboratory head. Scientific credit tends to concentrate on the "first author" (often a postdoc or student who ran the project) and the "last author" (usually the laboratory head). In many cases, "middle authors" (often "collaborators" from other laboratories) are coded as having "helped" by providing a "service," such as supplying biomaterials or performing specialized techniques.

In Mapping Lab, people recorded data in paper notebooks, managed dozens of samples in individual plastic tubes, pipetted liquids manually, and handled viscous radioactive liquids in plastic bags. To appreciate the painstaking nature of the work, consider an excerpt from field notes that describes a single step of one everyday procedure: the plasmid prep.[64]

The Beckman centrifuge had been spinning the samples at 45,000 revolutions per minute for four days, which separated the DNA into two bands. Maria carefully removed the plastic tubes containing the samples. "You cannot shake them," she said, explaining that agitating the liquid will cause the bands of DNA to diffuse, making it "impossible to pull out the DNA from the band we want." After slitting the top of each tube with a razor, Maria and I pierced each tube with a syringe, withdrew the lower band from each, discharged the fluid into a separate dialysis membrane, and sealed each membrane with clamps. After checking for leaks, we placed the membranes in TE solution. We planned to let them sit overnight and finish up the next day.[65]

Not only were laboratory techniques time consuming, but experiments sometimes failed to produce usable results. When they were successful, the evidence that they produced required careful visual inspection and interpretation.

In Mapping Laboratory, as in molecular biology more generally, the body of the skilled and conscientious individual was the central integrating technology for coordinating work.[66] People manipulated experimental materials using their eyes and hands. Bench space was individualized, and a shared understanding that no one should disturb the bench of another helped individuals maintain custody of their materials.[67] When materials were placed in common spaces, such as refrigerators, handwritten labels helped sustain a link between the individual and his or her "stuff." The "biographies" or "careers" of materials and inscriptions were carefully tracked. The operative knowledge-control regime delegated the task of creating a durable and auditable record of each experiment to individuals in the laboratory, who maintained a written notebook to record the flow of materials and inscriptions through complex, multistep protocols.[68]

Unlike some of the other laboratories where I did observations, Mapping Lab did not have a written "notebook policy." In another genome laboratory in 1989, an official document stated that all personnel should keep their records in notebooks with consecutively numbered pages; that "everything" should be entered, even mundane figures such as the amount of material weighed when preparing a solution; and that all records should be created contemporaneously, dated, and kept in ink. The policy also warned against writing notes on paper towels and argued that an accurate notebook can protect against a scientist's greatest danger, self-deception.[69] As in all sociotechnical systems, actual practice did not always match official instructions. In one laboratory, I observed a meeting in which a scientist and a technician spent hours trying to make sense of incomplete records. Such slippage in recordkeeping was regarded as an individual failure—in this case, that of the technician.

In Mapping Laboratory, a group of about six technicians formed the core of what people sometimes called the "assembly line" or "the factory." The laboratory head inflected these terms with a playful irony, a twinkle in the eye that seemed to underline the distance between their craft-work reality and the vanguard vision of high-throughput data production, involving hundreds of workers, robots, and machines in a highly automated manufacturing process. In contrast to the work given to technicians, the laboratory assigned postdoctoral associates and graduate students "projects" that were expected to result in publications. Most of these projects involved mapping but had objectives beyond producing map data and were not part of the assembly line. Even so, I heard complaints from some of the postdocs about spending time on "routine" mapping work rather than on career-enhancing projects and grumbling about doing the work of "mere technicians."[70]

Such complaints, along with maneuvering to boost morale, are familiar in molecular biology as scientists in training struggle to achieve independence.[71] But the perception that genome research posed special dangers to young scientists was acutely felt both by participants fully committed to the HGP and by scientists critical of the project. How would young scientists working on long-term, collective projects such as genome mapping stand out as individuals who had made notable contributions? Would their names be lost in long lists of coauthors when publications finally appeared? To molecular biologists, being part of a "routine" mapping project—even a prominent part—did not look like work that would allow a young scientist to produce the reputational resources needed to launch an independent academic career. Accordingly, the postdocs wanted to distinguish themselves from the pack by finding genes, developing new techniques, or unraveling biological mechanisms.

In Mapping Lab, progress often felt slow. The senior scientists and postdocs were fully committed to the vision of the HGP, but sequencing the human genome seemed a distant goal.

Pilot Sequencing Projects, circa 1991

As the HGP began, genome researchers launched a number of pilot sequencing projects. These projects, which included several efforts begun under the auspices of US genome program, aimed to sequence unprecedented amounts of DNA. Even more importantly, they sought to explore the challenges of large-scale sequencing and to develop methods. At Harvard, Walter Gilbert tried to make a new "multiplex" sequencing method into a practical production technology, which was tested by attempting to sequence *Mycoplasm capricolum*, a goat pathogen with a very small genome.[72] The notable projects also included the European project to sequence the genome of yeast; a collaboration to sequence the worm *C. elegans* launched by Robert Waterston of Washington University and John Sulston of the University of Cambridge in England; a Stanford University project to develop technology and sequence yeast; and the effort led by Leroy Hood of the California Institute of Technology to sequence a specific region of the human genome of importance in the immune system (the region of the T-cell receptor).

These pilot projects promised to produce a number of valuable outputs. Most obviously, they aimed to generate much longer continuous strips of sequence data than had ever been created. Genome researchers expected continuous sequences of bacteria and yeast to yield invaluable insights for the research communities that worked on these organisms and to provide the first detailed view of the overall structure of genomes. No one knew, for example, the number of genes found in a simple organism, such as bacteria, so scientists hoped that the pilot projects would yield insights about the basic machinery required for life.

Perhaps even more importantly, however, the genomics vanguard saw the pilot projects as a way to explore the challenges of large-scale sequencing and to experiment with technological and organizational strategies. Their goal was not to engineer "exotic" new sequencing technologies. Rather, they sought to imagine and implement new "sequencing strategies"—ways to speed production by deploying the available repertoire of technologies in new combinations, tweaking them, integrating them into informatics systems, and improving the flow of materials, data, and work. In this respect, the pilot projects were "experiments" for developing and evaluating sequencing strategies, and the data that a laboratory produced pertained not only to biology but also to evaluating the promise of specific sociotechnical systems.[73] More broadly, to the extent that some projects managed to sequence long, continuous stretches of DNA, they would yield something of enormous value to the nascent HGP: a demonstration that large-scale sequencing could be undertaken successfully.

Like genome mapping, sequencing at the beginning of the 1990s required meticulous, craft-style work, even in laboratories that used the first DNA-sequencing machines that Applied Biosystems Incorporated (ABI) and Pharmacia had recently introduced.[74] Laboratory workers used elaborate bench-top procedures to prepare material for sequencing: cloning and subcloning, purifying DNA, performing chemical reactions, and labeling DNA with radioactive or fluorescent tags.[75] To produce an accurate and continuous sequence of a region of the genome, the region had to be sequenced several times. The task of producing a "finished" sequence— integrating preliminary sequences into long, accurate strips—was a time-consuming process, complicated by such errors as insertions of extra bases, deletions of bases, or misidentifications of bases. Human beings had to "edit" the sequence, checking the original data wherever different "reads" disagreed in order to ascertain which one was correct. Additional problems stemmed from parts of genomes that proved difficult to sequence.[76] For example, strips of sequence that recur throughout the genome, known as "repeats," created ambiguities that made assembling an accurate sequence difficult.

The researchers who designed and managed these pilot projects operated in a complex and challenging world full of uncertainties, surprises, and opportunities. In the Matthews Laboratory—a US pilot project that later grew into a large operation—the scientists began making changes even before their project officially started, adjusting their sequencing strategy soon after completing their initial grant proposal in 1989. They also changed their plans for additional hiring, shifting from one postdoc and two technicians to three technicians. How the workforce would ultimately develop remained unclear, but they expected to employ both "senior technicians" and "factory-type" technicians once they moved into the "production phase." The laboratory also tested different sequencing machines, acquiring both an ABI and a Pharmacia machine to see which one worked better, as they put it, "in our hands."[77]

To manage the process of making incremental improvements, the Mathews Laboratory divided the sequencing process into five separate modules, so that "bottlenecks" could be addressed independently to increase overall throughput. One bottleneck that they eliminated early on was the need to compare and edit different "reads" by pouring over multiple paper printouts of sequences generated by the machines. An informatics specialist, MEG, wrote software that enabled them to line up multiple reads on a computer screen. The scientists also worked to simplify laboratory protocols to enable technicians with limited experience to perform them reliably. They were constantly looking for new ways to increase throughput—and not without reason. Their expressed goal was to produce 100,000 bases of sequence during year 1, scaling up to 400,000 bases in year 2.[78] Even at this scaled-up rate, sequencing the entire human genome seemed a daunting task. At 400,000 base pairs per year the job would take 7,500 years.

Sequencing Center, July 2000

Let us depart briefly from the loosely chronological structure of this book and fast-forward to the year 2000 to consider a rather different facility, supported by US government funds. SEQUENCING CENTER was a relatively small operation housed in an industrial park in an outer suburb of a major American city. In 1991, when I first met the scientist who later became Sequencing Center's director, he—like most HGP scientists at the time—was working to produce a rudimentary genome map, not to sequence the genome directly, and he defined himself as a genome "mapper" not a "sequencer." In the summer of 2000, the facility he directed, like other HGP laboratories, was engaged in the competition with Celera Genomics. The largest genome-sequencing centers then in operation, such as the United Kingdom's Sanger Centre and Celera, employed several hundred people. Sequencing Center was much smaller. In July 2000, it was a 5,000-square-foot facility with 30 full-time-equivalent personnel, making it roughly twice as big as Mapping Laboratory was in 1988 and quite a bit larger than the Matthews Laboratory during its first year. Twenty of Sequencing Center's employees were technicians, and a handful of PhDs, the center director, some computer programmers, and support staff rounded out the roster. The precise numbers shifted at the margins, mostly as technicians turned over, but the pattern of a few PhDs and many technicians was stable. There were robots and computers throughout the "wet laboratory" space, and the offices included an informatics room. Two rooms had banks of sequencing machines in them. One of these rooms resembled a traditional molecular biology laboratory with the usual benches fixed to the floor; the other was what the scientists called a "morphable" lab, with ventilation, electricity, and data cables all brought down from the ceiling to mobile stations that one could freely reposition by wheeling them around the room. The scientists thought of the morphable lab as the future. It would give them flexibility to reconfigure the space as new technology emerged.

Teams of technicians, grouped under head technicians, moved materials through the "data-production" process, prepping DNA, running sequencing reactions, and feeding the sequencing machines—a bank of ABI Prism 3700s, the new capillary gel machines that Perkin-Elmer's Applied Biosystems division had announced in 1998. The machines were collectively approaching 13,000 "reads" per week, with each read yielding some 600 to 700 base pairs of data. This level of "throughput" amounted to about 8 million base pairs per week—more than the entire accumulated holdings of GenBank in early 1988 and 80 times as much sequence information as the Matthews Laboratory aimed to produce during the entire first year of its pilot project. The machines automatically captured sequence data and exported them to the laboratory informatics system, which monitored production, tracking the flow of materials and information. Materials were kept in barcoded containers, and barcode readers were integrated into the informatics system. Workers scanned the barcodes of samples as they moved through the sequence-production process, and the system thus tracked the identity and location of each sample with minimal human intervention. A plate of DNA samples, of course, could "go missing" if, for example, someone accidentally dropped it on the floor. In such cases, the system would classify the plate as lost. The informatics system also monitored data quality, and if it detected problems, it would email the appropriate person. Because the laboratory spent $200,000 per week, it sought to discover problems quickly. A misaligned laser in one sequencing machine could cost $4,000 per day, mostly in wasted reagents and labor. The center director told me that the informatics system was like "Big Brother" because its intrusive eyes tracked everything, constantly searching for trouble spots. It sent senior staff daily emails reporting on the status of ongoing production. If the samples BILL handled on Tuesday turned out to yield unusually poor results, the system would discover this fact.[79]

To vary people's work and avoid the boredom of doing the same task day after day, Sequencing Center rotated technicians among "stations" according to a preplanned schedule. No one had a bench of his or her own; the "techs" moved among stations, each of which was dedicated to a different step in the process of sequencing. Describing a recent change in the organization of work groups, the center director explained that "this is all industrial sociology," and "it really has little to do with science." The center put considerable effort into building morale, even hiring a consulting firm to run a team-building retreat. Nevertheless, tensions existed in some work groups, especially within a group headed by a young woman who supervised several men her age or a bit older. When the men questioned her authority, her boss backed her up: "someone has got to be the foreman," he said.[80]

Much of the process of transforming the output of the ABI machines into completed sequence was handled computationally. Software assembled the individual "reads" into longer continuous pieces by finding overlaps. However, where ambiguities or gaps arose, more experienced laboratory personnel would do the hard work of producing "finished" sequence. These "finishers" would proceed through a series of troubleshooting protocols that often had to be adapted to the particular case. Finishing was performed at traditional benches and in many ways resembled the familiar craft work of molecular biology. The work of closing gaps often involved puzzle solving, requiring careful consideration of which protocols to apply and how to interpret the results.

Comparing Sequencing Center with Mapping Lab and the Matthews Lab reveals profound change in roughly a decade—and not only in hardware, size, and the amount of data produced. If the Mapping Lab of 1988 relied mainly on craft-based techniques, the Sequencing Center of the year 2000 applied late twentieth-century manufacturing methods to the mass production of information, with the informatics system, not the body of the individual person, playing the central integrating role. Sequencing Center operated under a knowledge-control regime that little resembled that of a molecular biology laboratory—a fact evident in different modes of territorializing space, organizing workflows and the division of labor, and imposing control and accountability on personnel. The technicians and scientists' embodied skills remained important, but the task of choreographing the production process, monitoring it, and tracking the flow of materials and inscriptions had been shifted from the individual to a distributed socio-technical system integrated by an informatics system. Computers captured data automatically and coordinated the creation of an auditable trail of records. Checking the quality of data, once a task that required skilled interpretive work, had been automated to a great extent. Protocols had been redesigned to simplify work, and modes of processing materials had been redesigned to utilize robotics. The technology of data production and the knowledge-control regime governing the internal operation of the laboratory had been thoroughly transformed.

* * *

This brief look at a sequencing facility as the HGP drew to a close provides a picture of the nature of the changes that occurred in the internal socio-technical structure of genome laboratories during the 1990s.[81] With these dramatic changes in mind, let us return to the time when sequencing the human genome still remained a distant goal and factorylike sequencing facilities existed only in the imaginations of the genomics vanguard.

3   Laboratories of Control

Scientific laboratories not only produce knowledge objects but also regulate the status of these objects in knowledge-control regimes. This chapter examines regimes and practices that control the transfer of knowledge objects into and out of laboratories, focusing on genome mapping and gene hunting at the outset of the HGP. How did the knowledge-control regimes governing these laboratories frame the problem of regulating these transfers? Through what practices did actors attempt to establish and resist this control? How did these regimes and practices articulate with the knowledge-production process? And in what ways, if any, was the regulation of transfers of knowledge objects implicated in shaping the *objects* themselves?

In addressing these questions, I seek not only to develop an account of how knowledge and control are coproduced in laboratories but also to lay the groundwork for analyzing how knowledge-control regimes changed during the HGP. Members of the genomics vanguard wanted both to create transformative knowledge and technology and to achieve unprecedented sharing of data and materials. Given the complexity of sequencing the human genome, they expected the project to require new levels of trans-laboratory cooperation. In the chapters ahead, we will examine how they tried to realize these changes, but first we must establish a "baseline" by considering the prevailing modes of control in molecular biology and human genetics at the outset of the project. This chapter accordingly focuses on knowledge control in laboratories involved in genome mapping and human genetics from 1988 to 1992.[1]

At the end of the 1980s, human genetics and genome mapping both remained grounded in the small-laboratory culture and bench-top technology of molecular biology, although genomics was beginning to differentiate into a new form of science. US genome centers at the time also looked like slightly scaled-up versions of traditional molecular biology laboratories, resembling MAPPING LAB much more than the factorylike data-production facilities that later emerged, such as SEQUENCING CENTER. The main goals of *genome*-mapping laboratories (constructing genetic and physical maps) differed from the goals of *human genetics* laboratories (hunting for specific disease genes). However, because even incomplete genome maps could greatly speed a gene hunt, the boundary separating the two activities was unstable and elusive. Most human genome mappers participated in gene hunts—although building maps, not searching for genes, was their main goal. Human genome mapping thus was entangled in the intense competition of gene hunting. In light of these similarities and interconnections, I sometimes treat genome-mapping and human genetics laboratories as belonging to a single category: genetics and genomics laboratories.

The initially liminal differences between "genetics" and "genomics" laboratories grew more pronounced as the HGP got underway. Genome centers began working to build high-throughput data-production systems, and some of them, such as EASTERN GENOME CENTER, soon succeeded in doing so. New knowledge-control regimes for governing large-scale mapping and sequencing were also taking shape, as we will see in chapter 4. Change was unevenly distributed but, at the outset of the genome project, the prevailing practices for regulating access to data, materials, and other forms of knowledge tended to be similar to what the vanguard believed that genomics needed to leave behind.

The chapter begins with a discussion of the regime of the laboratory. It then analyzes control over knowledge objects in several contexts: conversations, requests for data and materials, scientific meetings, and multilaboratory collaborations.

The Regime of the Laboratory

At the outset of the HGP, most genetics and genomics laboratories (such as Mapping Lab) resembled the molecular biology laboratories studied by Karin Knorr Cetina.[2] One can readily draw on her analysis to produce an account of their governing frame. This frame parses the world into two spaces—inside and outside of the laboratory—granting the laboratory head managerial privileges over the operation of internal space and over the entry and exit of knowledge objects. The laboratory head is entitled to direct the research program, select and manage personnel, allocate projects to subordinates, authorize the expenditure of resources, and make internally authoritative judgments about the epistemic quality of knowledge objects. The laboratory head also enjoys strong managerial privileges over transfers of knowledge and resources and holds a legitimate monopoly on representing the laboratory and its accomplishments to the wider world. All of these managerial privileges can be delegated, but authority ultimately resides in the laboratory head.

To be sure, the laboratory head's control is not absolute. He or she is obliged to conform to government regulations and organizational policies—for example, regarding health, safety, human subjects, and financial accounting. Moreover, junior scientists enjoy some legitimate entitlements, such as a moral claim to receive credit, mainly in the form of coauthorship, for "their" contributions to "their" projects. Even so, the laboratory head enjoys much discretion to establish the boundaries of projects and to allocate credit according to his or her perceptions. Junior scientists therefore often worry that they will not receive the credit that they deserve. Overall, the governing frame treats the laboratory as the lab head's dominion, a jurisdiction under his or her exclusive control.[3]

Here I call the knowledge objects found in the laboratory its "holdings." These holdings typically include not only devices, materials, machines, and documents but also techniques, knowledge embodied in persons, and information about the laboratory's activities and plans. Holdings are not isolated objects; they are woven together into evolving assemblages, and many of them undergo significant changes in form and meaning as laboratory work proceeds. They are also of extremely unequal "strategic value," a term I mean to encompass a number of often intertwined forms of value— for example, epistemic, technological, political, exchange, and use value (as perceived by the actors). From the start of the HGP, genome scientists paid a great deal of attention to the strategic value of their holdings as well as to others' holdings, assessing the "promise" of a clone, the "power" of a technique, the "excitement" of a finding, the "practical import" of an analysis. What unifies the concept of strategic value is a future-oriented outlook focused on what an object will (perhaps!) enable an actor to *do*. Assessments of strategic value are typically based on some combination of hard-nosed analysis, experience, imagination, and affect. Like other knowledge claims, judgments about strategic value have a collective dimension; they may become a matter of widespread agreement, or, especially at the frontiers of research, they may be contested.[4]

Like other circulating things, holdings have "careers" or "biographies" that trace their flow through material, social, and discursive space.[5] For the laboratory head, choreographing these flows is an important strategic matter. Which knowledge objects should the laboratory seek to export and import? When, how, and under what terms and conditions? And what should it seek to contain in the laboratory and through what means? Well-orchestrated transfers—both inbound and outbound—are essential to create opportunities, stake knowledge claims, build a track record, acquire resources, and maintain a position at the forefront of research.

Transfers rarely involve mere changes in the physical location of objects; they also entail changes in control relationships—as occur, for example, when findings are transferred from the dominion of the laboratory into the published literature. In human genetics at the time of the HGP, as in many other fields, the incentive to publish "major findings" quickly was strong, but during a race to find a gene, publishing "intermediate" results could aid rivals, who might use those results to find the gene and win the lion's share of the credit.[6] The laurels bestowed on the gene hunter for "progress toward" finding a disease gene paled by comparison to those showered on the laboratory that finds the gene. Yet it was often necessary to publish "progress-toward papers" for purposes of accountability. In gene hunting, strategic questions about whether and when to publish intermediate results arose regularly. Analogous issues surrounded more targeted transfers, such as providing clones or other valuable objects to specific colleagues.

Scientists used various practices to exploit their holdings strategically. One practice widely used by academic researchers as well as by those operating in explicitly corporate settings was to restrict access to valuable holdings and use them to generate additional knowledge that they could "cash in" later. Another strategy was to carefully target access; for example, unique biomaterials might help entice potential collaborators who held complementary knowledge objects. Another was to export holdings—say, via publication in the literature or conference presentations—and thus to build reputational resources. The question was often not simply which strategy to undertake but how best to utilize multiple strategies in combination.

Timing could be key. Waiting might offer short-term advantages, and mistakes could damage reputation, but speed was also valuable. The question of whether a holding was "ready" for transfer and to what jurisdiction could be a complex one.

Strategic considerations also were important regarding inbound transfers. Genome researchers eagerly sought out emerging knowledge and were often delighted when it accelerated research or provided powerful insights. Researchers constantly scanned the literature, attended meetings, and talked to colleagues to learn about potentially interesting developments, and they traded gossip on the research and plans of colleagues and competitors. Nevertheless, given the challenges of acquiring, assessing, and incorporating novel knowledge objects into laboratory practice, deciding which new knowledge objects to import—and how much to invest in them—was tricky, especially when evidence of each one's value was incomplete.

The Perimeter of the Laboratory

The jurisdiction of the laboratory head—like jurisdictions found in many other knowledge-control regimes—is not a neatly delimited physical space but a virtual region constructed through ongoing practices.[7] For analytic purposes, we can think of this virtual region as bounded by what I call a "perimeter." This term was chosen less for its meaning in geometry than for its usage in police work, where "establishing a perimeter" entails the dynamic management of an emphatically breachable boundary. The perimeter of the laboratory is not coterminous with its physical architecture. To be sure, spatial arrangements remain relevant; Thomas F. Gieryn and others have shown how architecture matters in biotechnology and how walls and locks—some of the oldest and most important technologies of privacy—play a significant role in protecting laboratories and their contents.[8] But as Steven Shapin demonstrates in the case of Robert Boyle's "house of experiment," the threshold of a laboratory is made less of stone than of social convention.[9]

The perimeter of the laboratory is constituted—and on occasion breached—using a variety of means, and practices aimed at regulating access to the holdings of laboratories play a central role in science. To begin analyzing how this virtual region is constituted in ongoing action, let us make two crude, initial distinctions: the first between "restricted" holdings (which the laboratory seeks to keep in a confidential space) and "unrestricted" holdings (which the laboratory allows others to access); the second between "released" holdings (which the laboratory allows to travel "outside" its physical threshold) and "unreleased" holdings (which the laboratory keeps "inside").[10] Setting aside momentarily the fact that there are degrees of both restriction and release, we can identify four types of holdings—positions in a two-by-two table. *Restricted, unreleased holdings* might include such entities as secrets, plans, emerging results, work in progress, and unique materials protected by keeping them within the physical confines of the laboratory. *Unrestricted, released holdings* (which could originally have been produced in the laboratory or elsewhere) might include, say, information gleaned from published papers or scientific meetings, data posted on the Internet, commercially available products, and DNA samples available from open repositories. *Unrestricted, unreleased holdings* would include entities—perhaps software or details of experimental protocols—that the laboratory would, if approached, happily make available to others. The final category, *restricted, released* holdings, might initially seem to be an oxymoron, but the protected space that a laboratory constructs can extend far beyond its physical confines. For one thing, actors can construct "protective containers" designed to enclose objects even as they circulate outside the laboratory walls. These containers are produced using a diversity of machinery, including the relatively tangible (such as sealed packages) and the relatively virtual (such as encrypted digital files). Other mechanisms include those that rely on trust (such as a colleague's promise) or sanctions (such as penalties for violating a nondisclosure agreement). Absent leaks (an ever-present possibility), restricted, released holdings never cross the perimeter of the laboratory—even as the perimeter is "stretched" to reach new physical and social domains.

In addition to packaging holdings in protective containers, laboratories can inscribe freedom and constraint directly and materially into the holdings themselves. During the normal course of laboratory work, materials and information undergo many transformations for a variety of technological and epistemic reasons.[11] Some of these transformations are irreversible operations that render their outputs unsuitable for particular purposes. These transformations are often ordinary parts of scientific practice as mundane as performing statistical operations—for example, taking the arithmetic mean of a set of numbers—that cannot be reversed to yield the original data. Irreversible transformations thus create *new* knowledge objects with different uses and value. But they do more: they simultaneously create opportunities for laboratories to build constraint and freedom of action directly into the very knowledge objects that they choose to circulate. Routine procedures for extracting the DNA from biological tissue entail an irreversible transformation that laboratories can deploy to enforce control relationships because the extraction process destroys the cells in which the DNA is found. Giving a colleague a small sample of purified DNA restricts his or her future actions by limiting the number of experiments that can be undertaken with the sample. In contrast, providing live cells enables the recipient to obtain more DNA simply by growing more cells. Providing live cells thus yields more control to the recipient.

Irreversible transformations also include changes introduced for the express purpose of imposing constraints. The example of redacting a document is particularly instructive. The term *redaction* normally connotes the concealment of segments of a document that has already been committed to writing.[12] However, there is no reason to neglect what one might call "preemptive redaction"—the removal of information during the writing process itself. Scientists (like many other authors) often carefully choose to leave certain things unsaid, and omissions are sometimes a crucial *part of* a scientific text.[13] The transformations associated with writing and editing, along with those arising from aggregating and manipulating data or from purifying and packaging biomaterials, create means through which a laboratory can build its perimeter directly into its products. Indeed, the flexibility with which assemblages of data, materials, and documents can be selectively disassembled and/or bundled together to build constraints into knowledge objects makes the concept of *restricted, released holdings*—which initially seemed like an oxymoron—into a category that encompasses many—perhaps most—of the objects that laboratories export.[14]

Controlling Transfers and Transferring Control

So far I have discussed in general terms how genetics and genomics laboratories managed the selective transfer of knowledge objects at the outset of the HGP. Let us now turn to specific examples, selected from three modes of interlaboratory communication: face-to-face interaction, requests for data and materials, and presentations at scientific meetings.[15] My goal is to qualitatively analyze the complexities that arise in each case, not to make strong quantitative claims about prevalence.[16]

Face-to-Face Interaction

The perimeter of the laboratory is constructed on many occasions, including during specific face-to-face encounters between laboratory members and outsiders. In such situations, knowledge is selectively and simultaneously revealed and concealed through informal speech.[17] As one would expect, genome scientists often told each other about preliminary findings, answered detailed questions about techniques, and discussed current work and future plans. These collegial interactions, often grounded in ongoing relationships with familiar persons, provided opportunities to request and receive advice or to describe problems and results. Even among strangers, such shoptalk was treated as a normal courtesy, consistent with the broader climate of intellectual exchange. Short-term visitors to laboratories were often told and shown a great deal.[18] In 1993, I accompanied an American scientist on an extended tour of a French genome laboratory, during which a lab scientist spent hours providing thorough answers to the American's highly technical questions.

Researchers also engaged in strategic information control during such conversations, attempting to restrict the transfer of specific bits of valuable knowledge during conversation as they, in effect, constructed the perimeter of the laboratory by regulating speech.[19] Consider a conversation that that took place during a social gathering at an academic laboratory.

Pat, the lab head, called out loudly: "Wei, you have to listen to this." Wei, a postdoc, had recently created some unique and valuable DNA probes in the genomic region near IMPORTANT DISEASE GENE. Pat continued, now addressing the small group of postdocs, students, and technicians: "Wei, we have to be especially careful what you say about the work we are doing. Wei, be very careful because when competitors want to get information, they don't ask the senior people questions. Often the students and the postdocs," Pat continued, "will let out the lab secrets without even knowing that they did. There are very sly ways of asking questions. All of you have to develop ways of not answering the questions that you are not ready to answer."

Pat then said that Wei and RICH, another postdoc, were going to a meeting that competitors would attend. "So we've developed a strategy for that meeting. Rich has been given instructions about what he can and cannot say to them. If anyone pushes him too hard, he's supposed to say, 'Pat said that I can't talk about that result yet and that you have to call if you want to know any more about it.'"

Everyone had listened attentively to Pat, but at about this point LIL, another postdoc, joked that MARIO, who was visiting and expected to return to LOCATION soon, had to stay permanently at the laboratory, saying: "You have too many secrets, Mario. You can't leave."

"No one is to talk about my probes!" Wei said. But after this statement, which I took as expressing real concern, he started joking, "I don't want to tell Rich any more information about my work since he is going to too many meetings." A technician, JIM, joined in the banter, pointing to Rich, who had recently accepted job at OTHER LABORATORY: "Rich is going to come back to visit, and no one is going to talk to him. He's going to be saying, 'Hey, what's going on? No one will even say good morning to me.'"[20]

This episode illustrates one way that laboratory heads aimed to constitute the perimeter of the laboratory by regulating speech to control the exit of valuable knowledge objects (such as the very fact that Wei's markers existed). It also raises the question of how far the effort to contain laboratory secrets should go. By playfully pointing out the extremes to which a policy of containing secrets could conceivably be pushed, the joking raised questions about the limits of efforts to tightly control the perimeter.[21] The absurd examples—not allowing Mario to leave or refusing to say good morning to Rich—suggested that overzealous control could interfere with "normal" interaction in this "academic" setting.[22]

Pat was not the only laboratory leader whom I observed instructing subordinate personnel not to mention specific holdings outside of the lab, and many scientists also asked me to treat some information as secret.[23] Leaks, of course, remain an ever-present possibility, particularly as the numbers of people involved grow and where interpersonal ties transect the laboratory boundaries.

*   *   *

Let us now turn to an audience's perspective on conversations about a laboratory's activities and resources. Audience members often cannot discern how much the flow of knowledge objects is being constricted. They typically cannot observe precisely the location of the perimeter of the laboratory and may under- or overestimate the amount of authorized disclosure or the amount of leakage. Nor can they definitively ascertain the reasons for any suspected concealment. The following conversation, which took place during a break at a Cold Spring Harbor meeting in 1991, illustrates these multiple layers of uncertainty.

I was walking out of the auditorium with ANDY, a molecular geneticist, when he called out to two scientists he knew, BILL and CLYDE, with a follow-up question. They stopped, and after a minute Andy asked Bill, who is from UNNAMED UNIVERSITY and has connections to COMPANY ONE, "Is COMPANY ONE going to do anything with capillary gel electrophoresis?" Bill kind of smiled coyly in a way that suggested that this question was treading near the boundaries of confidential territory, but he said with what I read to be an "I can't really tell you, but let me tell you anyway" tone that he would be very surprised, extremely surprised, if they didn't do anything with it. Clyde started asking more technical questions. He seemed to know that some of the people involved in the COMPANY ONE/UNNAMED UNIVERSITY nexus had been studying capillary gels. The issue that Clyde considered important was modeling the location of DNA molecules over time as they move through the gel, taking account of the effect of different fluorescent tags. A mathematical understanding of these differences is necessary for sequencing machines because the algorithms that "read" the sequence must take account of them. Clyde continued probing Bill with technical questions until Bill finally said, "Yep, the gel mathematics, they are now understood." Clyde replied, "I never saw the reference." Bill explained, "COMPANY ONE did the study, and COMPANY TWO has also done the study." Clyde said, "Oh, that's why." No one said the obvious—namely, that the companies had decided not to publish in order to retain a competitive edge in the race to improve sequencing machines. It was clear that Bill would say nothing more, and no one pushed him further.[24]

This vignette illustrates a complex dialectic of revelation and concealment as information is concealed, revealed, and concealed as it is revealed. The scientists appear to self-consciously recognize their efforts to control the flow of information, and I experienced the conversation as one in which the unsaid seemed to speak in portentous yet ambiguous ways. In particular, Bill, an academic scientist, seems to be artfully releasing information he learned from Company One (that the company now understands the gel mathematics) while withholding other information (even the briefest gloss on the findings).[25] At the same time, Andy and Clyde seem interested in extracting information from Bill. The artful construction of backstage regions—by Bill and (apparently) by both companies—is a salient aspect of the interaction. In addition, everyone seems to accept that for the company to impose a regime of secrecy in this context is well within the normal range of action.

The conversation also illustrates the complex texture of information control because it features several types of secrecy and multiple layers of selective disclosure. The unpublished findings of research on gel mathematics that the two companies apparently conducted seem to fall into the category that the national security community calls "objective secrets," a kind of knowledge imagined to be unchanging and universally valid, like an equation needed to design a thermonuclear weapon. Bill's statements also involve what that community terms "subjective secrets"— information of the "loose lips sink ships" variety that is related to a specific, transient situation.[26] His comportment and speech seem to self-consciously point to a boundary separating permissible from impermissible speech about Company One's activities, although (in what may have been another layer of strategic information control) he took no steps to clarify whether he was edging up against that boundary or breaching it while coyly maintaining deniability. It is also possible that to deter competition or to encourage exuberance among stock analysts and their technical consultants (Could Clyde be such a person?), Company One wants relevant people to "know" that it understands the gel mathematics yet does not want to tell outsiders details. Bill thus seems to invite the other participants to *know that* the company knows but without permitting them to *know what* the company knows (or does not know).

On one level, this conversation, though brief, involves knowledge-control practices pertaining to both "objective" secrets (about gels) and "subjective" secrets (about company activities). More deeply, however, the conversation points to ambiguities in the distinction between objective and subjective secrets. Company One's unpublished findings about gel mathematics (which might be imagined to be objective secrets) remain, at least for outsiders not privy to the work, incompletely vetted and consequently of questionable epistemic status. Although company insiders might have extensive grounds for being convinced, outsiders cannot be sure. Bill seems to believe that the phenomenon is "now understood," but

should one share his confidence? Can one be certain that Bill is not bluff-
ing or overstating his certainty? Should Bill's assertion that Company
Two "has also done the study" be taken to indicate that the two companies have done "the [*same*] study" (something Bill does not exactly say)
and have found similar results? Can we assume that the two (possibly)
secret studies "replicate" each other? Company outsiders who implicitly
or explicitly answer "no" to such questions will not perceive the company's knowledge of gel mathematics to be a scientifically certified objective
secret. Indeed, by adjusting their level of skepticism, outsiders can slide
back and forth from "we know that the company knows the objective
secrets of gel mathematics but will not disclose them" to "we know only
that one presumably knowledgeable quasi-insider has disclosed the subjective and possibly secret claim that the company has studied and understands gel mathematics but hasn't published what that person says the
company knows."

As this example suggests, the knowledge-control practices effected in
such conversations shape not only the social distribution of objective and
subjective information but also the social distribution of the capacity to be
confident about which information can be regarded as subjective or objective. Far from only being about access, as consequential as that is, practices
that enact the perimeters of laboratories also shape the epistemic qualities
of knowledge.[27] Multiple boundaries—of what is considered reliable knowledge, of who has access to what knowledge, of who knows how much about
what knowledge is being (selectively) contained—are simultaneously
constructed. More broadly, this discussion shows how knowledge-control
regimes predicated on tight containment trigger proliferating epistemic
constraints.

Requests for Data and Materials

Let us turn to interlaboratory requests for data and materials. A first step
is to note how the regime of publication in scientific journals constitutes
such requests. The governing frame of this regime treats requests for
the data and materials underlying published papers as normal and legitimate, imposing on authors of scientific papers the duty to provide access
to the data and materials, such as clones, that are needed to verify or build
on the authors' published work. Qualified scientists have a correlative
right to access. To be sure, the regime allows the impracticality of granting

access to certain kinds of data and materials to temper this overarching rule.[28] Nevertheless, in what this regime treats as the ordinary case, the burdens and entitlements remain quite clear. Duties imposed by the publication regime thus curtail the managerial privileges that the laboratory regime grants the laboratory head regarding transfers. Similar cross-regime curtailments are found at many points where the jurisdictions of knowledge-control regimes overlap.

In genetics and genomics laboratories at the outset of the HGP, postpublication requests for data or materials were common. Scientists often requested DNA samples (such as clones and cell lines) used in published work. These requests were often executed through a simple exchange of letters: for example, an initial letter from the requester to the source identifying the biomaterial and usually saying something about the experiments that the requester intended to undertake, followed by a reply from the source to the requester providing the material and perhaps including some helpful technical information.[29] By convention, the requester would generally acknowledge the source in publications using the materials. Some mapping projects gathered materials from many laboratories. The group that published the first physical map of a human chromosome (the tiny Y chromosome) thanked more than 60 scientists for "human DNA, blood, or cell line samples."[30]

At times, negotiations placed constraints on how the requester could use materials. Requesters of clones might agree not to pass them on to anyone else and typically would briefly outline their research plans. Such statements of intentions were not necessarily considered binding, however, given the contingencies of research. As BETH put it, "Usually no one has said that [one's initial plan] is the only thing you can do, and anyway that's very hard to, that's difficult, because your experiment can change a little bit. ... I always say what I want the clone for. But I don't think [that] once I have the clone, I'm really restricted to [doing just] that."[31] More extensive negotiations, or at least full disclosure, might be deemed necessary when the requester anticipated direct competition with the supplier. In such a case, Beth felt an obligation to "ask them for permission to do the same experiment here" and sent a letter "exactly describing" what was planned.[32] In this individualistic and contingent world, restraints on use of transferred clones were grounded mainly in informal and potentially flexible etiquette.

Although laboratories often granted requests for materials and data, I also heard countless stories of requests being ignored, addressed slowly, or explicitly refused, perhaps with an explanation (e.g., the material is in "short supply"). Sources sometimes promised to send materials, but the materials never actually arrived. For the person holding a resource, the disincentives of complying with requests could be significant; at a minimum, they included the hassle of preparing materials for shipping, which sometimes was not trivial.[33] In situations featuring focused competition, such as a race to find a gene, the risk of aiding rivals loomed large. Not surprisingly, requesters often attributed delays and denials to competition and duplicity. One scientist, ERIN, described her frustration in trying to gather materials for a new project in human genome research: "There were few labs like MENDOLA'S LAB that had a lot of clones. ... He would say 'yes,' and I would never get them. ... He would say, 'Let me talk with my collaborators, and I'll get back to you.' Then he'd never get back. He'd go, 'I haven't talked to this collaborator.' It goes on and on and on. I mean, I can deal with 'no.'"[34] Later, one of the collaborators told Erin that Mendola never mentioned her request, saying that had he done so, she would readily have agreed to provide the clones. Some scientists thus honored the rules of the publication regime in their verbal formulations, if not in their actions.

The frustrations of having requests ignored led some scientists to pursue a strategy of pushy persistence. One scientist described following the initial request with a second letter: "*In case my letter or your reply got lost in the mail, eight weeks ago, requesting your clone such and such, as described in such and such a paper, etc., etc.*" A third letter would eventually follow: "*In case my last two letters, or your last two replies, got lost, etc., etc.*" The cost was only the price of the stamps, and perhaps the holder of the material would get embarrassed and send the clone.

Scientists sometimes privately offered justifications for the seemingly endless delays that requesters reported. In addition to the time required to answer requests or the limited supply of material, some researchers also argued that their postdoctoral associates or graduate students, who had worked hard searching for a gene, deserved an opportunity to receive career-enhancing credit that might be lost if another laboratory used their materials to win the race for a gene.

*   *   *

Up to this point, we have focused on requests for data and materials underlying published papers—requests that authors, at least according to the governing frame of the publication regime, are expected to honor because many published results cannot be replicated or built upon without access to such materials as clones. Laboratory heads are also sometimes asked for data and materials that are not associated with published papers, but because these materials remain outside the jurisdiction of the publication regime, the researcher is not obliged to grant such requests. Clearly, the laboratory head has the privilege of making prepublication transfers, but even if he or she wishes in principle to do so, strategic, epistemic, and practical considerations may arise.

An example from MARIGOLD LABORATORY, a research group known for its openness, illustrates some of the considerations involved in prepublication transfers.[35] In the early 1990s, Marigold constructed a groundbreaking genome map and published several notable papers based on this work. The first paper described a significant result—one that was achieved using an unpublished version of that map. The paper also announced that Marigold had created the map and that the map would be described in a subsequent publication. In effect, Marigold divided an assemblage of data and materials into two parts: the *result*, which Marigold transferred to the publication regime via the initial paper, and the *map*, which the laboratory did not circulate pending preparation of a second paper. The perimeter of the laboratory was thus built directly into the initial paper, which simultaneously kept one knowledge object (the map) inside the laboratory while publishing two others (the result and the fact the map existed).

Immediately after Marigold's paper announced the existence of the map, scientists began to request access to it. Marigold scientists expressed a clear desire to distribute it widely as soon as possible, but they believed that the map, which was still in "preliminary" form, was not yet ready.[36] JON, the postdoctoral associate who was running the project, was busily polishing the map, which he said was "usable" in his hands but needed to be rechecked before being broadly released. Marigold was "obligated to make sure that what goes out is true," Jon stressed, so the laboratory would release the map only after "we've done all the error checking we can possibly do, and ... we are confident in it." Even so, he explained, they had already made some data available to a few other laboratories. In one case, a group of scientists was working on a chromosome that Jon had already carefully checked, so he gave them the map of that chromosome. In another instance, the requesting laboratory had enough experience to "understands the caveats" of unverified data. Jon explained that "basically we kind of feel them out when they call" to ascertain how much they know about the technical details of this kind of work. The requesters who called were also trying to find out if Marigold would make the map available to them" later. "They don't want to go to the expense of setting up experiments unless they'll know that THE MAP is available."[37]

Genetics and genomics laboratories—even those my interlocutors described as "open"—thus displayed caution when crafting papers and managing prepublication requests, and they took account of a mixture of strategic and epistemic matters as they fashioned knowledge objects for transfer to other regimes and jurisdictions. Such cases nicely illustrate a more general point: creating delimited knowledge objects, managing their transfer between knowledge-control regimes, and constructing the perimeter of the laboratory are inseparable aspects of a single process.

Presentations at Meetings

Laboratories also selectively release knowledge via presentations at scientific meetings. At the Cold Spring Harbor meeting, the premiere annual meeting of HGP scientists, researchers presented their work either in plenary-session talks or in poster sessions.[38] Almost all talks and posters were coauthored yet presented by a single individual. Presentations at these meetings posed opportunities and risks, and senior scientists reviewed drafts of abstracts and posters by postdocs and graduate students and coached them, sometimes right up to the last minute, to regulate epistemic claims and sometimes to manage interlaboratory relations by discouraging impolitic remarks.

Among the most delicate questions that laboratory heads faced when crafting presentations was deciding how much unpublished knowledge to disclose. Describing work close to the research front was necessary to reach beyond the already familiar, but overclaiming was important to avoid, as was providing unreciprocated help to competitors in situations of focused competition, such as races to find genes. When the HGP began, searching for a gene entailed building detailed local maps of the region in which the gene was believed to reside. In many cases, competitors could immediately employ local map data from other researchers, perhaps combining it with their own data to improve maps or narrow the search. Researchers therefore often perceived a strong disincentive to disclose knowledge relevant to a gene hunt, and more than a few of my interlocutors described human genetics meetings as games in which the players silently negotiated about how close to keep their cards to their chests. Genome scientists—especially those who worked on nonhuman organisms, such as the worm, the fruit fly, or microorganisms—often described human genetics as particularly secretive. The worm community, a small community that could trace its lineage to the laboratory of Sydney Brenner, was perceived as especially "open" and "collegial."[39]

A pattern of tight regulation of outbound transfers was evident at the international Single Chromosome Workshops (SCWs), a series of small meetings, each focused on a different human chromosome. The SCWs brought together dozens of leading gene hunters and genome mappers (as well as those involved in both activities) to exchange data and develop state-of-the-art maps.[40] Gathering together data and materials was essential to the SCW goal of constructing state-of-the-art chromosome maps. Collecting the most recent information in the Genome Data Base was equally important—to the success of this new database and to making genome data broadly accessible. However, many SCW participants were hunting for the same genes, so they faced contradictory pressures to release data and to hold it closely. Most of the time, questions about the availability of knowledge objects remained submerged below the surface of interaction, though sometimes they became visible or were explicitly discussed.[41]

Researchers might not reach decisions about how much data to present and how much to hold back until after the meeting began. At one workshop, Carol told me she had not decided whether to present her latest data: "There are a few problems, so it's preliminary data. It needs to be checked out, but I may decide to present it. It depends on what my competitors do. If the competition presents incomplete data, then I will present incomplete data. I'm going last, so I'll get to decide." When she gave her talk, she *did* present the preliminary results.[42]

The SCWs featured working sessions in which, as one meeting organizer put it, the "real work" of assembling a map gets done. At one meeting, I observed some two dozen people from North America, Europe, and Japan spend hours compiling and comparing data YAC clones from the chromosome:[43]

The process proceeded deliberately: "OK," the chair would say, "we'll move on to chromosomal region S13." After giving the name of a YAC from the region, he would ask: "Does anyone have a size of this YAC?" The scientists would compare sizes and note any discrepancies. People occasionally mentioned YACs from sources other than the widely circulated libraries. In one instance, no one had any YACs at a particular location except for a group from NAME OF COUNTRY, who had three such YACs. "Are they available?" the chair asked. The answer: "Maybe in a couple of months." No explanation of the delay was offered, and no one pressed the issue.[44]

In such settings, many scientists appeared to be quite forthcoming, although generally audiences could not tell whether data were being quietly withheld. Researchers believed that delayed release (that is, temporary restriction of access) was common. The noncommittal statement "maybe in a couple of months" is an example of what we might call "announced" delayed release because both the existence of the object and the delay were disclosed. Unannounced delay, in contrast, is typically not detectable, although my interlocutors sometimes made skeptical comments about the paucity of results emerging from other research groups, suggesting that they "must" have some results "by now." Although delays might have stemmed from technical problems or a desire to recheck data, people typically suspected other motives.

Another strategy was "data isolation"—that is, providing access to isolated portions of an assemblage of knowledge objects that had been taken apart in ways that made it difficult or impossible for other laboratories to reconstruct or build on results. A variety of transformations offered ways to constitute knowledge objects that could be released in this isolated sense. One technique was to rename a YAC or other clone that was in one of the widely distributed libraries, thus obscuring its identity. Some laboratories gave clones aliases that they used internally (in part to prevent leaks), but the use of nonstandard names in scientific meetings was controversial. Critics angrily pointed out that the only reason to rename publicly available clones was to make them hard to recognize, and at one workshop a well-known scientist from COASTAL GENOME CENTER told a plenary session that people who renamed clones deserve "to be shot."[45]

Participants in the SCWs remained somewhat skeptical about promises that resources would be made available in the future. Much of the excitement at one SCW concerned a new collection of especially promising YACs developed by a EUROPEAN GROUP. The group told workshop participants that this library would soon be distributed. KATHLEEN, a postdoc at a genome center, expressed doubts about the timing, suggesting that a longer delay seemed likely. She said that it looked like they did want to make the library widely available, but it would be "in their rights" to keep it for themselves. But to say so at this meeting "just wouldn't do."[46]

Seeking to encourage more "sharing" of data, the European and American funding agencies that supported the SCWs sought to change the rules of participation in the workshops. After some participants held back data at several early SCWs, policy makers began seriously discussing the imposition of formal "data-sharing" requirements on SCW attendees. These policy makers, of course, recognized that gene hunters were cagey with valuable resources, but they wanted to crack down on those who behaved as "spectators," collecting data but contributing none; presenting data at the meeting but never submitting them to the Genome Data Base; or presenting findings but declining to release the materials needed to verify or extend the results.

At a policy meeting in 1992, the funding agencies proposed that SCW participants be required to make all data and materials presented available immediately. The proposal encountered objections from many of the gene hunters present. Nevertheless, a guideline was adopted soon thereafter, requiring that "any data and materials" presented "are automatically considered to be immediately publicly available."[47] These guidelines, however, constituted a weak knowledge-control regime. Beyond the reputational risks that gene hunters had often proved willing to accept, the SCWs' only available sanction was to deny travel reimbursement. The policy may have eliminated some spectators and curbed data isolation, but it predictably did not solve the problem of data release. In particular, researchers could still resort to unannounced delayed release, and, not surprisingly, some decided to do so. One example was ULRIKE, a German who was part of a three-way race to find the gene that causes a specific BRAIN DISORDER. At an SCW held in 1993 under the new rules, she told me that she had carefully avoided mentioning her most important new results—four extremely promising clones from the region of the gene—because she did not want to give them to her competitors: PIERRE, also a participant in the meeting, and RICHARD, an American who had been expected to attend but did not show up.

The SCWs proved unable to impose a tight knowledge-control regime. In the context of zero-sum races to find genes, the policy could not override the incentives for delayed release, and the autonomous scientist's managerial privileges could not be easily curtailed. Indeed, the policy seems to have led some scientists to actually reduce their level of disclosure at future SCWs by inhibiting announcements of resources that they had planned to make available in the future.

Collaboration Regimes

Not only did the laboratories I studied control transfers by managing conversations, presentations, and requests, but without exception they also participated in what they described as "collaborations." These exchanges were indispensable to remaining at the cutting edge of research, but they often entailed tricky negotiations about control. Genome scientists used the term *collaboration* to refer to a broad range of multilaboratory projects.[48] Most frequently, the term described projects expected to result in coauthored scientific papers, but it also applied to situations in which only an acknowledgment was anticipated.[49] The term also designated agreements between commercial firms and academic laboratories as well as interfirm alliances and joint ventures among firms. (Commercial agreements are not discussed in this chapter; some prominent ones are addressed in chapter 5.)

For our purposes, "collaboration" can be understood as a genre of knowledge-control regime based on an agreement among specific agents to participate in some joint research project or activities. As such, collaborations have an ad hoc quality, taking many shapes. For example, they vary significantly in the extent to which they are seen as a durable organizational unit. Some collaborations last and evolve for many years.[50] Others are one-off deals centered on the transfer of a single knowledge object. The nature of the interaction, the kind of work, the sorts of resources exchanged, the intensity and duration of the project, and the extent of legal or quasi-legal formality all vary.[51] Control over the joint activities that collaborations constitute need not be equal, and rights in the knowledge objects emerging from collaborations are often unevenly distributed among participants, with terms tailored to the situation. Despite this variability, however, it is possible to describe a generic governing frame that captures the most basic elements and structure of this category of regime:

- The *domain* of the collaboration: an imagined joint research project or activity with (sometimes loosely) defined boundaries.
- The *parties* in the collaboration—that is, two (or more) agents with the managerial privileges needed to make commitments.
- The *resources* each party commits to the collaboration—for example, specialized expertise, materials, technological capacity, personnel, data, money, and so on.
- *Expectations about the progression of action*—for example, about the trajectory of research activities, the flow of data and materials, the division of work, and the results to be produced.
- *Expectations about governance*—that is, about control relationships pertaining to resources, to decision making, and to entitlements and burdens during the work process and regarding the results produced.
- A *mutual agreement* to enter into the collaboration that expresses (at least apparent) alignment of how the parties imagine the other elements of the frame.

This schematic frame, obviously the most generic of templates, follows the basic structure of a contract among autonomous parties. As such, it lacks substance until it is filled in with specific agents, projects, resources, and expectations. In practice, agreements to collaborate may be verbal or written. They vary considerably in the extent to which they explicitly elaborate details about scope, duration, distribution of credit, and allocation of managerial control rather than rely on implicit—and presumably mutual—understandings.

Constructing Collaborations

In the HGP, collaborations often formed when two or more laboratories possessed potentially complementary resources.[52] Prospective collaborations typically identified a domain-specific project and relevant resources that the parties possessed. Many considerations might affect the attractiveness of a potential collaboration, including judgments about the scientific merit of the project, the likelihood it would succeed, the resources each party would provide, the value of those resources, the strategic position of the different players, and the allocation of authority and credit. Asymmetries among the parties—in resources or in the importance of the collaboration to each party's overall research effort—often influenced whether an agreement was reached and, if so, what its terms would be. Genome researchers frequently expressed the idea that collaborations should be mutually beneficial and that rewards should match contributions. This moral vision, although uncontroversial in the abstract, became a point of contention in situations where one party deemed the other to be expecting too much or delivering too little. Researchers also expressed the idea that the "best" and most satisfying collaborations required exchanges of skills, judgment, expertise, or ideas, not just what they coded as "technical" "services."

Collaborations were built around a variety of resources. Laboratories that developed new techniques or significantly improved old ones sometimes found themselves inundated with requests to collaborate. A single example will suffice:

Rob, the head of the laboratory, described how collaborators constantly approached his laboratory. Usually the collaborators would have some probes that they wanted his group to RUN THROUGH THEIR SYSTEM. His lab would get acknowledged for having produced the data or, if it was a longer-term project, would get coauthorship. They had a lot of requests and were very selective, and with insufficiently interesting projects he would tell the person to forget it. But if somebody called up and said, "Rob, I think we've found the gene for such-and-such," or "We've got a candidate gene," then he would say yes to the request right away.

Collaborations also formed around strategically valuable biological samples. As an example, consider how Silke, a German postdoc in an American laboratory, found a pathologist to supply her with the tissue samples that she needed to compare the DNA in cancer cells with normal cells from the same patient. She and the pathologist reached a verbal agreement that he would give her the material and be named as an author on the papers. In addition to the patient cells, Silke's work required a battery of specially selected probes to identify changes in the tumor DNA. These, however, she did not gather herself. Her supervisor had a good collection of such probes, mostly obtained from other laboratories. The suppliers of the probes would get an acknowledgment in any papers that Silke published, and she would also cite their original paper describing the probe. Unlike the pathologist, they would not receive coauthorship of Silke's papers.

Silke's collaborations illustrate how granting coauthorship and acknowledgments provided a means to award several levels of credit. Because the value of coauthorship could be diluted if there were too many authors, researchers tried to keep the length of author lists to a "reasonable" num-ber—for example, by using acknowledgments and citations rather than coauthorship to credit the providers of probes. Efforts to balance the ten-sion between awarding credit and diluting credit in a manner everyone regarded as "fair" were only sometimes successful, and disagreement about credit arose regularly both before collaborations were fully formed and after they began to produce results.

* * *

During fieldwork, I rarely directly observed negotiations about the forma-tion of collaborations, not least because the settings in which collabora-tions take shape—phone calls, side conversations at meetings—were sporadic and elusive. However, the instances in which I did observe frag-ments of such negotiations suggested that the strategic dimensions of col-laborations came up regularly. In one case, I was allowed to overhear one side of an extended phone call in which a EUROPEAN GENOME MAPPER discussed a multilaboratory collaboration with a GENOME SEQUENCER who was being invited to join one of the teams racing to find an important disease gene. The proposed arrangement was for the mapper to send the sequencer some relevant DNA to sequence. The conversation included a systematic evalua-tion of the competing teams in the race as well as plans to have the most visible representative of the multilaboratory team send a letter to the sequencer's main funder stressing the importance of the project. Ultimately, the sequencer did join the collaboration.

On occasion, I also observed the members of a single laboratory assess-ing the promise of alternative research strategies, some of which would involve entering into collaborations. Consider a meeting at the NELSON LAB held shortly after a postdoctoral associate had returned from a small scien-tific meeting.

The postdoc began by presenting the latest news about the likely location of DEADLY DISEASE GENE. The news was mixed: good because scientists had established the gene was in a particular region, bad because that region was large enough to contain perhaps 100 genes, so finding that particular gene would be an arduous task. The postdoc then presented nine different experimental schemes for honing in on the gene. The laboratory head and the other scientists assessed each one.

"This is brute force."

"There has to be another way."

"That's really grasping at straws."

One more promising idea required collaborating with Randolph, who could screen his YAC collection to find YACs from that region. The Nelson Lab could use those YACs to zero in on the gene.

"Yes," Sheila said, "but Randolph has said that if anyone sends them samples and they screen their collection, then he is going to release those YACs to the public." This was presented as a downside because competing laboratories would also get the result.

"That's OK," Nelson replied.

A bit later Nelson presented a tenth idea, a collaborative "scheme that we've been invited to participate in." He then outlined the proposed collaboration, which relied on a technique consistent with his laboratory's core expertise.

As in this case, evaluations of potential research strategies can mix "technical" questions (How likely is the scheme to succeed?) with questions about competition and collaboration (Is it OK for competitors to get the YACs?).

*   *   *

Matters of compatibility and trust also figured into my interlocutors' thinking when assessing prospective collaborations.[53] Some of the most productive collaborations in the field were intertwined with personal friendships. Patrick, the Southern Genome Center director, explained his outlook on personal relationships in collaboration in what I regarded as unduly modest terms: "It's primarily a choice on the basis of the science, but I can tell you that there is a second filter of what it's like to work with people. Really strong scientists, in my view, judge what they do only by the science that is involved. I respect that. I think it's the right decision. I can't function that way."[54]

Preexisting animosities sometimes impeded the formation of collaborations. In one instance, Jon, a postdoctoral associate, conducted a worldwide search and identified a particular collection of DNA samples as the best material for his project. He wanted to negotiate access to the collection, but Michael, a senior scientist, had a long history of competing with the group that controlled it. "Michael didn't want to have anything to do with the group at any cost," Jon said. "My insistence was that I had seen the data and that I didn't care about what their personal conflicts were. ... Scientifically it was logical." After several months, Michael finally relented.[55]

To assess the trustworthiness of potential collaborators, some scientists actively sought out "gossip" and "scuttlebutt," drawing on the networks of interpersonal connections. When prospective collaborators began approaching Kathleen about accessing a strategically valuable DNA collection, she asked some colleagues at a Cold Spring Harbor meeting about their experiences working with them. During an interview a few months later, she emphasized the importance of reputation in decisions about potential collaborators:[56]

Kathleen: [Whether the agreements need to be] formal or informal depends on who you're dealing with and what their reputation is, and it's very political. [Laughs]
SH: Yes. And certainly some people in the community have the reputation for somebody who you've got to be real careful in dealing with.
Kathleen: That's right, and you'd better lay out very specifically what you expect.
SH: Would there be another category even beyond that who you wouldn't—
Kathleen: Would not collaborate with? Yes.
SH: Wouldn't even enter into—
Kathleen: Yes. [Laughs] Wouldn't even consider it.[57]

We can read such efforts to evaluate reputation not only as self-protection but also as a mode of collectively sanctioning unreasonable behavior. However, asymmetries of power—along with the post hoc nature, uneven distribution, and questionable credibility of scuttlebutt—limit the effectiveness of reputational sanctions.

Structural Fragility

Like all research, collaborations run into contingencies. Many are deemed successful, some become unexpectedly productive, and some prove to be fragile. At times, the recalcitrance of materials and techniques leads scientists to drop or redefine projects. In addition, the structure of collaborations—as ad hoc agreements to conduct joint work in a hyperindividuated culture—regularly leads them to fall apart, whether from mismatched expectations or from other asymmetries. I heard many stories about collaborators who fell short of their partner's expectations, and I sometimes observed researchers questioning their collaborators, a few times very angrily, about how work on their "end" was proceeding. Many stories also circulated regarding expectations about authorship that were not met. One scientist succinctly summarized the variety of surprises that collaborators sometimes experienced when papers came out: "You had an acknowledgment when you thought you were going to get an authorship. Or you thought you were going to get an acknowledgment, and you got nothing. Or you thought you were going to get an authorship, and you got nothing. Or ... you got an authorship, and you never knew that you were going to get authorship, and you never had a chance to read the paper."[58]

The contingencies of research and the challenges of cooperation in a hyperindividuated culture can interact, making it difficult to maintain mutual understandings about the domain, resources, temporal progression, and governance of a collaboration.[59] Tensions about the pace of progress also arise when collaborations are more important to one partner than to the other, leading to uneven expenditures of resources. Materials and techniques also may fail to perform, and when things do not go as imagined, terms may have to be renegotiated. When Kathleen encountered a situation in which her collaborator's materials "didn't pan out," she decided to renegotiate terms: "Pretty much the person who has the resource, which in this case is me, can dictate the terms. If the other people don't like those terms, then they don't have to collaborate anymore."[60]

Ambiguities also may develop about the boundaries of a collaboration. Understood as a jurisdiction under some form of joint control, the collaboration has a boundary, defined (often loosely) in terms of its substantive and temporal reach. Even if this boundary is clearly defined in the initial agreement, it may grow indistinct as work proceeds. The endpoint may become ambiguous because research results often provoke scientific questions that arguably fall both inside and outside the domain as originally conceived. Disputed endpoints sometimes develop not only in collaborations among different laboratories but also when junior scientists move to new laboratories. Bitter disputes may arise about who has the privilege of conducting follow-up work: the scientist who is departing from the laboratory or the head of the laboratory where the research began.[61]

At times, the very existence of a collaboration, as opposed to a competitive relationship, became ambiguous owing to the lack of a well-formed governing frame. In some cases, for example, initial overtures about possible collaboration were met not with outright rejection but with vague promises of future exchanges of data, leaving the situation unclear to one or more of the parties. In one instance, EUROPEAN GROUP provided the same set of valuable DNA samples to two other groups, US GROUP and CONTINENTAL GROUP, to use to search for the same gene. According to US Group's account of events, US Group and Continental Group initially agreed to exchange data back and forth. The two groups worked in parallel, not actively exchanging data, although US Group expected that they would do so in the future. Soon US Group was making great headway, a fact that became widely known after the laboratory head presented preliminary results at a scientific meeting. Continental Group then proceeded to submit a paper for publication without telling US Group in advance. When European Group learned of the submission, it informed US Group that a paper was in the works. US Group then stepped up its efforts, moving additional people to the project in what became a feverish race to fully document the discovery and publish first. In this way, vague discussions produced an ambiguous relationship that was supplanted by intense competition. Such shifts are probably especially likely to occur when researchers get close to reaching a goal collectively defined as an important end point, such as identifying a disease gene.[62]

The ad hoc nature of collaborations and their susceptibility to asymmetrical expectations often set the stage for breakdowns and contestations of control. Moreover, a knowledge-control regime based on ad hoc agreements between autonomous parties offers only the thinnest of governing frames, providing no standardized template for reaching settlements about control relationships or jurisdictional boundaries. The absence of a means for arbitrating disputes or effectively sanctioning those who violate agreements contributes further to their instability. In many cases, ambiguities about terms, disagreements about the value of knowledge objects, or tensions surrounding allocations of resources and authority make collaborations hard to create and difficult to sustain, especially when the trajectory of research diverges from initial expectations.

Conclusion

The epistemic problem of constructing knowledge and the sociopolitical one of constructing control over knowledge are tightly intertwined. Knowledge-control regimes, such as the publication regime and the laboratory regime, provide generic frames for interpreting and guiding action, simultaneously imposing order and creating room for maneuvering. In this account, laboratories emerge not merely as sites of knowledge production or simply as participants in economic or quasi-economic exchange but as sites in which control relationships are built—quite tangibly—into knowledge objects, agents, spaces, and the relationships that tie them together. Irreversible transformations and circulating restrictions, for example, allow the perimeter of the laboratory to be redrawn in ways that shape the (in)capacities of both knowledge objects and their users.

Knowledge-control regimes often grant certain agents, such as the laboratory head, strong control over a particular jurisdiction, but these entitlements may encounter resistance and sometimes utterly fail. Confidentiality may be breached by loose talk or deliberate leaks. Policies mandating data sharing may be circumvented or backfire. Collaborations may collapse owing to the fragility of the ad hoc regimes sustaining them. An ongoing process of dynamic maneuvering, especially at the boundaries of regimes and their key categories, produces specific configurations of knowledge and control relationships.

To many members of the genomics vanguard, the practices in genome-mapping and human genetics laboratories that slowed the flow of data and materials posed a serious problem for genome research. Competition needed to be channeled in productive directions, they believed, and the field needed to move beyond the distrust and tight containment of knowledge objects typical of those fields at the end of the 1980s. How, then, could the genomics vanguard constitute knowledge-control regimes capable of sustaining new, more collaborative forms of genome science? Examining their strategies for accomplishing this goal requires an entire chapter: the next.

4   Research Programs and Communities

The rise of a self-consciously revolutionary research program, such as the Human Genome Project, can create a context in which the settlements that govern a scientific arena become especially susceptible to renegotiation. The realm of what seems possible, so often narrowed by the naturalization of existing modes of governance, expands for a time, inspiring efforts to restructure the terms of interaction among the participants in the world of research. From such moments, new knowledge-control regimes sometimes emerge, perhaps contributing to durable change.

This chapter examines just this sort of moment through a comparative analysis of two knowledge-control regimes that factions of the genomics vanguard built at the outset of the HGP. In the United States, the decision to move ahead with a $3 billion genome project raised urgent questions about how to govern this nascent and novel form of biological research. A regime for governing the US genome program emerged at the end of the 1980s and early 1990s, and it became an important resource in ongoing efforts to maintain order in genome research and the HGP. Here I compare the US regime with the Reference Library System (RLS), which also took shape during the late 1980s and the early 1990s in the Genome Analysis Laboratory at the Imperial Cancer Research Fund (ICRF), a London-based charity.[1] Neither regime was codified in a founding document resembling a formally written constitution—although documents defining control relationships were significant in both cases. Nevertheless, each regime established a coherent governing frame, and each aimed to constitute order in genome research and in relations between genome research and broader scientific communities in genetics and molecular biology.

The fate of these two knowledge-control regimes differed greatly, however. The RLS vision failed to capture the hearts and minds of most science policy makers, and it never received anything approaching the large-scale funding that the regime established by the US genome program enjoyed.[2] In keeping with the methodological principle of symmetry, comparing a successful regime with a less successful one provides analytic leverage. Even so, as the undertaking of a single laboratory, lacking the formal standing of a national program, the RLS might seem at first glance to be too different from the well-funded US genome program to serve as the basis for a fruitful comparison. But a closer look reveals the limitations of this view. First, comparing the US regime with the RLS reveals structural and normative choices embedded in both of these regimes that might be inadvertently naturalized if only one regime were analyzed. Second, as I show, the fact that a regime of the form instantiated in the RLS never won massive government funds is not an independent variable, unrelated to questions of regime structure. The leaders of the US genome program were well aware of the RLS and *actively rejected* its vision, when in principle they could have adopted its approach and showered it with funds. We therefore cannot treat the RLS's smaller budget and its lack of official standing as an explanation of its fate. As we will see, comparative analysis helps to explain why US policy makers rejected the vision that the RLS advanced and illuminates some of the challenges that vanguards (or factions thereof) face when trying to introduce new knowledge-control regimes.

My account begins with the US regime and then turns to the RLS, considering several questions: What kinds of agents did each regime constitute? How were rights, duties, and other control relationships distributed among those agents? What forms of accountability did each regime seek to create, channel, and check? How did each regime interweave specific technologies and modes of control? And what forms of resistance did each encounter? The conclusion examines the vision of the scientific community that each regime sought to advance and offers an explanation of the two regimes' different fates.

Building a Regime for the US Program

In the late 1980s, the proponents of mapping and sequencing the human genome framed it as a project that necessitated a large-scale and carefully managed program. This frame played a central role in justifying a concerted mapping and sequencing effort, but it also raised questions of governance. Existing regimes for funding molecular biology, institutional-ized in such funding agencies as the NIH Institute of General Medical Sciences, were wholly committed to the figure of the autonomous scien-tist—an individual laboratory head who submits grant proposals reflecting his or her interests. To the genomics vanguard, a regime that made a virtue of funding individual grants without any centralized planning, targeting, or coordination seemed incapable of managing the HGP. One genome project leader, OLLIE, likened an attempt to run the HGP in this manner to what might have happened if NASA had approached the Apollo program in the same way:

"Well," [NASA officials would say,] "we have this plan to go to the moon, and three times a year we will accept applications for activities which you think might help us get there. The sole basis on which we will decide which of these to fund is just which ones seem most exciting to various little panels that we will assemble." Off in the corner will be some guy saying, "There really is going to have to be a guidance sys-tem in this rocket, and we don't seem to be getting many applications in this area." At this moment, the apparatchiks would say, "Not to worry, we've got these very exciting applications here on fuel, and we're going to fund a lot of those because everyone agrees they're very exciting."[3]

The situation raised questions for the genomics vanguard. How could the US genome program be properly governed and kept on track? If the project entailed building large genome centers, how would those centers interact with and relate to the small laboratories typical of molecular biology? What new control relationships and accountability mechanisms would emerge, and to whom would these modes of discipline apply? When biologists who were not part of the genomics vanguard asked simi-lar questions, they often grew worried. Might the arrival of "big biology" tie up funds and restrict the autonomy of the individual scientist? Would large genome centers come to dominate all of molecular genetics? In this context, the genomics vanguard self-consciously recognized that it needed to assure other biologists that the project would not destroy their way of life.

Defining the Scope of the Project

By 1988, when the US Congress committed to funding a genome project, the governing frame of a knowledge-control regime for the US program had begun to take shape.[4] Most significantly, the policy-making process had constituted the program as a semiotic and bureaucratic entity and defined its parameters by

1. *Delimiting its scope* with a set of long-term goals (e.g., mapping and ultimately sequencing the genome of the human and several model organisms; making this "data set" available in public databases; and developing technologies for genome analysis).[5]

2. *Specifying its technical focus and rationale,* laying out long-term objectives and a vision of why they were worth achieving, and putting some areas of genetics, such as hunting for disease genes or studying human diversity, outside its mandate. The US program would enable these kinds of research but would not center on them. Its raison d'être would be to *produce* data and technology for the entire scientific community to *use*; that is, the ultimate goal was framed almost as a "service" activity of generating data that would benefit all biologists.[6] The vision of revolutionary change, along with the significant technoscientific challenges, attracted some extremely creative researchers who expected the project to lead to an exhilarating scientific future.

3. *Positioning it in organizational space,* ostensibly under the joint control of the newly formed National Center for Human Genome Research (NCHGR) within the NIH and the Office of Biological and Environmental Research (OBER) within the DOE. In practice, the NIH soon became the unofficial lead agency, with the DOE serving as a junior partner.

4. *Articulating budgetary expectations and temporal boundaries* ($3 billion over 15 years). In 1989, NIH program head James D. Watson defined a period within which expenditures could be calculated and progress evaluated, announcing an "official" starting date (1990) and a target "completion" date (2005).

This governing frame proved durable. As one might expect, ongoing work was required to maintain it—for example, by shoring up the boundaries of legitimate uses of genome program funds.[7] Moreover, at times some HGP leaders argued that the NIH and DOE were failing to take program goals seriously enough and run a truly tight ship. Ultimately, however, the basic elements of this governing frame remained in place throughout the project, and they provided the criteria according to which the funding agencies later pronounced that the HGP had reached its goals "ahead of schedule" and "under budget."[8]

The science policy process during the late 1980s also brought into being an informal but influential group that one might call the "leadership" of the US program—a network composed of members of the genomics vanguard, with the NCHGR and (to a lesser extent) OBER at the center. This network linked NCHGR and OBER officials to outside science advisers and to directors of important genome laboratories. The funding agencies, backed by this network of advisers, became the agents with ultimate responsibility for managing the project as a whole. This network included scientists—most of them specialists in genetics, gene mapping, DNA sequencing, informatics, and computer science and mathematics—from major research universities, medical schools, and the DOE's national laboratories. It operated through both informal consultation and formal advisory committees. Outside advisers also included some experts in bioethics, law, and social sciences to advise the ELSI program. Demographically, the leadership was disproportionally white in comparison with the US population. Most of its members were men.[9] Indeed, men held almost all of the higher leadership positions (e.g., as funding-agency directors, laboratory heads, and so on) throughout the HGP.

Beyond constituting the HGP as an entity with a mission, a budget, and a leadership, the initial governing frame also defined another important agent: "genome centers"—large laboratories dedicated specifically to accomplishing HGP goals. Early advocates of the HGP had imagined constructing relatively large, multidisciplinary genome laboratories in order to yield economies of scale, produce synergistic effects, and advance cutting-edge genome science. The success of the HGP as a whole would significantly depend on these laboratories. To concentrate resources in one or two huge centers seemed unwise and unworkable: unwise because investing in just a few centralized facilities could lead to a spectacular failure, and unworkable because concentrating resources to that degree would be politically unpopular, even—or perhaps *especially*—among some of the project's most ardent supporters, many of whom wanted to play a major role in the HGP. The DOE had already launched three genome centers at its national laboratories in 1987, and in public statements the NIH envisioned funding 10 to 20 genome centers. The NIH planned to issue requests for proposals and to evaluate proposed centers using its advisory and peer review mechanisms.

Building New Control Relationships

Defining the project and winning a place for it in the science-funding bureaucracy was a significant achievement. But when Congress committed to the project in 1988, crucial aspects of the governing frame remained to be articulated. The loose modes of control of molecular biology, with its reliance on uncoordinated investigator-initiated grants, did not allow for long-term planning and ongoing coordination, but a new regime had not fully taken shape. One strategy for tightening managerial control over the project was to establish intermediate goals to guide research and impose accountability. Toward this end, the NIH and DOE constituted a planning group of outside scientific advisers and personnel from each agency to develop a joint "five-year plan." The plan took shape in the summer and fall of 1989 and was published in 1990.[10] It formulated a set of goals, some of which were quite specific and expressed in quantitative terms (box 4.1).

Even more importantly, the planning process helped to institutionalize the principle that the funding agencies, especially the NIH, would have extensive managerial privileges over the project. The NCHGR had issued to the scientific community its first Request for Applications (RFA) for genome center grants in July 1989, but the five-year-planning group rejected that RFA, arguing that it envisioned genome centers that were too unfocused. The planning group wanted tighter control over genome centers, fearing that if left to their own devices they might drift off task whenever they encountered something more immediately "interesting" than mapping and sequencing, such as studying biological mechanisms or hunting disease genes. With their specific objectives inadequately articulated, centers might flounder and squander resources. Worse, unproductive centers might grow entrenched and hard to "kill."[11] In an unusual move, the NCHGR withdrew its initial RFA before three full months had passed, announcing that a revised one had "superceded" it.[12] The new RFA emphasized accountability, stressing that genome centers should be tightly focused on achieving specific genome project goals. Scientists who sought to establish genome centers could propose their own strategies for reaching these goals, but the funding agencies would intensely evaluate their plans and progress: "Each center must have tangible and, where possible, quantifiable aims that define a specific goal that the center intended to accomplish during the granting period. The center will be accountable for the attainment of such

Box 4.1
NIH–DOE Five-Year Goals, 1990

1. Mapping and Sequencing the Human Genome

   Genetic Mapping

   - Complete a fully connected human genetic map with markers spaced an average of 2 to 5 cM [centimorgans] apart. Identify each marker by a sequence tagged site (STS).

   Physical Mapping

   - Assemble STS maps of all human chromosomes with the goal of having markers spaced at approximately 100,000-bp [base-pair] intervals.
   - Generate overlapping sets of cloned DNA or closely spaced unambiguously ordered markers with continuity over lengths of 2 Mb [megabases] for large parts of the human genome.

   DNA Sequencing

   - Improve current and develop new methods for DNA sequencing that will allow large-scale sequencing of DNA at a cost of $0.50 per base pair.
   - Determine the sequence of an aggregate of 10 Mb of human DNA in large continuous stretches in the course of technology development and validation.

2. Model Organisms

   - Prepare a mouse genome genetic map based on DNA markers. Start physical mapping on one or two chromosomes.
   - Sequence an aggregate of about 20 Mb of DNA from a variety of model organisms, focusing on stretches that are 1 Mb long, in the course of developing and validating new and improved DNA sequencing technology.

3. Informatics—Data Collection and Analysis

   - Develop effective software and database designs to support large-scale mapping and sequencing projects.
   - Create database tools that provide easy access to up-to-date physical mapping, genetic mapping, chromosome mapping, and sequencing information and [that] allow ready comparison of the data in these several data sets.
   - Develop algorithms and analytical tools that can be used in the interpretation of genomic information.

4. Ethical, Legal, and Social Considerations

  • Develop programs directed toward understanding the ethical, legal, and
  social implications of Human Genome Project data. Identify and define
  the major issues and develop initial policy options to address them.

5. Research Training

  • Support research training of pre- and postdoctoral fellows starting in
  FY 1990 [fiscal year 1990]. Increase the number of trainees supported until
  a steady state of about 600 per year is reached by the fifth year.

  • Examine the need for other types of research training in the next year
  (FY 1991).

6. Technology Development

  • Support automated instrumentation and innovative and high-risk
  technological developments as well as improvements in current
  technology to meet the needs of the genome project as a whole.

7. Technology Transfer

  • Enhance the already close working relationships with industry.

  • Encourage and facilitate the transfer of technologies and of medically
  important information to the medical community.

Source: This synopsis appeared in DOE 1992, 5.

milestones through yearly progress reports, an annual center directors [*sic*]
meeting and the competitive renewal process."[13]

The revised RFA emphasized that to keep centers "focused" and to assure
that they make "sufficient progress," frequent scientific reviews would be
required. NIH staff would not only review annual progress reports and hold
meetings of center directors but also fund centers for "an initial term of five
years." As the genome program reached its initial goals, "the focus of the
HGP Centers and of individual grants will change." Centers could not
assume that their funding would be renewed. An evaluation would take
place after three years, with the idea that the final two years would allow
centers that received unfavorable reviews "sufficient time for submission
and review of a revised [renewal] application" or "for orderly phase-out of
the grant."[14]

Setting Standards

The NCHGR further tightened the accountability mechanisms for governing genome centers by setting standards intended to make progress

unambiguous and quantifiable. To HGP leaders, establishing standards seemed particularly urgent in genome mapping. Evaluating the quality of a map was difficult because scientists lacked a basis for comparing different types of maps. Each type—genetic linkage maps, restriction maps, radiation hybrid maps, and contig maps—offered a distinct view of the genome.[15] Further complicating assessment, different laboratories often used different technological strategies to produce the same kinds of maps. Given the uncertainties about which mapping methods would work, the leadership encouraged experimentation and a diversity of technological strategies. But at the end of the day, it wanted to evaluate outcomes.

During the preparation of the five-year plan, the advisory group developed a proposal to require all mapping centers to report data using a particular type of mapping "landmark"—which they dubbed a "sequence-tagged site" (STS). In September 1989, four genome project leaders outlined the concept in a short position paper in *Science* informally known as the "STS proposal."[16] An STS landmark consists of a short string of sequence data that occurs only once in the genome, producing a unique site identifiable by that sequence—much as a phrase that appears only once in a book can serve as a searchable landmark indicating a specific location in the text. The STS proposal argued that creating a standard landmark would provide a "common language" for translating among different kinds of maps, facilitating the merger and comparison of different maps and offering a means to check quality and measure progress. Requiring centers to translate their maps into STS form thus promised to simplify the task of evaluating the relative efficiency of different technological strategies and the performance of different laboratories.

The STS standard also was intended to guarantee that the wider scientific community would be able to access and use genome maps. Previous maps depended on landmarks based on clones, and the use of such a map required physical possession of the clones. Without the appropriate clones, for example, one could not take a random piece of human DNA and determine its location in the genome. This reliance on clones—biological materials that had to be carefully stored and shipped to map users—made these genome maps quite different from, say, highway maps, which can circulate in printed or electronic form. A clone-based genome map in written form was of very limited use; the clones, in effect, were an indispensable *part* of the map.

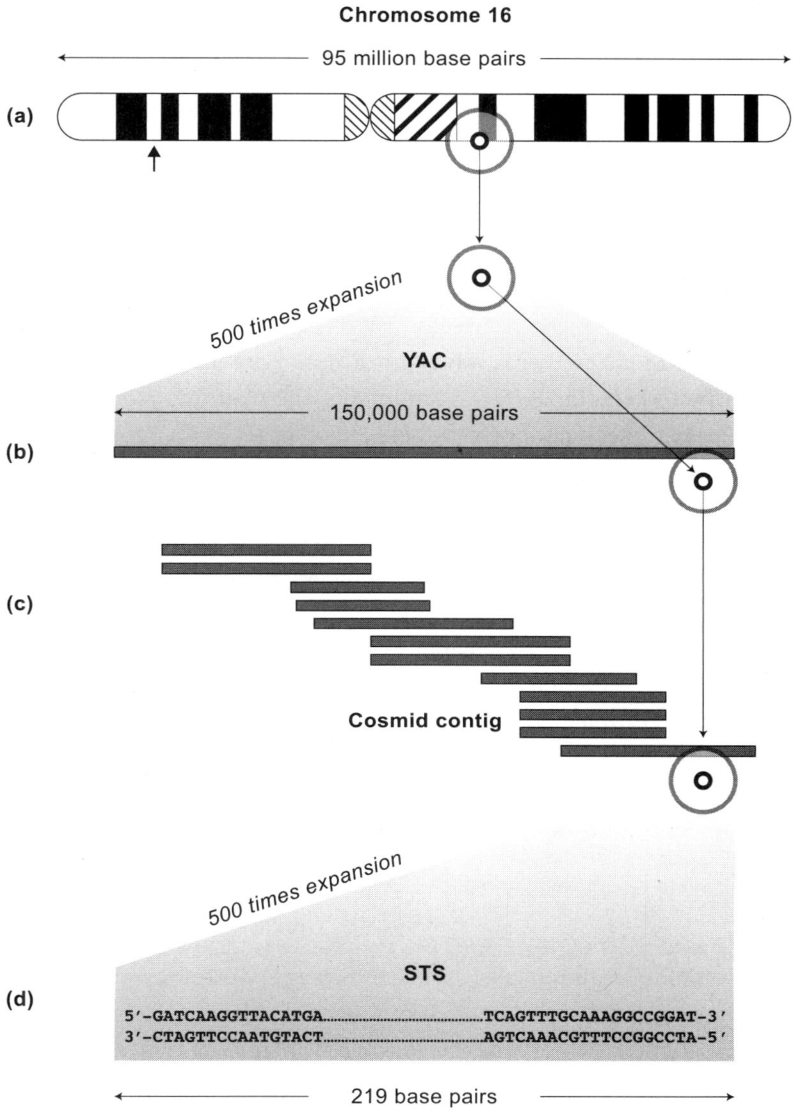

Chromosome 16

← ———————— 95 million base pairs ———————— →

(a)

500 times expansion

YAC

← —————— 150,000 base pairs —————— →

(b)

(c)

Cosmid contig

500 times expansion

STS

(d)

5′–GATCAAGGTTACATGA..............................TCAGTTTGCAAAGGCCGGAT–3′
3′–CTAGTTCCAATGTACT..............................AGTCAAACGTTTCCGGCCTA–5′

← ———————— 219 base pairs ———————— →

Figure 4.1

Connecting maps using STSs. During the early 1990s, genome mappers began using STSs—unique landmarks detectable by PCR—to link maps at different levels of resolution. The location of these STSs on a variety of physical maps could then be determined. The example given here, from a project to map chromosome 16, shows an STS **(d)** that was localized using in situ hybridization to a 3 to 4 million base-pair region on the map **(a)**. It was also localized on a 150,000 base-pair YAC **(b)** and on a cosmid contig **(c)**. In addition, polymorphic STSs could also serve as genetic markers, allowing them to connect genetic linkage maps to physical maps. *Source*: Adapted from Doggett, Stallings, Hildebrand, et al. 1992, 204–205. Illustration by Chris Cooley, Cooley Creative, LLC.

at some of these very expensive projects like the Name of Institution one, and they looked unreviewable from the outside," he explained, "because they were completely tied to this clone collection." The only way that another genome center could ever really ever tell "whether these maps were any good" would be to get huge numbers of these clones and repeat the work. The "initial driving force" behind the STS proposal was "to get the system open" by making it relatively easy to "grab any little part of anybody's map and see whether it was right or not."[20] By making maps more transportable and combinable, STS landmarks, the planning group hoped, would make genome maps easier to share and verify.

The US genome program framed its five-year goals for physical mapping in terms of numbers of STS landmarks to produce: 30,000 STSs an average of 100,000 base pairs apart.[21] By requiring that genome centers publish maps using STS landmarks, the funding agencies aimed to achieve technological goals and impose specific control relationships simultaneously. In effect, the STS standard expressed and at the same time tightened a governing frame that cast genome centers in the role of *producers* and *distributors* of genome data, with a duty to get high-quality data into the hands of the scientific community quickly. "Ordinary" molecular biology laboratories were cast as unconstrained *users* of genome maps, free to employ them for any purpose without upfront negotiations or encumbrances on their freedom of action.[22]

Monitoring Productivity and Progress

The accountability requirements of this knowledge-control regime encouraged genome centers to increase their productivity. These centers' interests were mostly aligned: they shared the goal of advancing the HGP, which still had vocal critics, and were not racing to map the same DNA because each center was mapping a different region of the genome. In human mapping, the centers were organized by chromosome. The center at the University of California, San Francisco, for example, aimed to map human chromosome 4; the Lawrence Livermore Laboratory sought to map chromosome 19; and the center at the Whitehead Institute at the Massachusetts Institute of Technology (MIT) set its sights on mapping the entire genome of the mouse. They made resources available to one another—for example, by circulating libraries of clones, software, and know-how.[23]

Even so, genome mapping remained quite competitive. Rather than the "focused competition" of gene hunting, mapping was mostly an area of "diffuse competition," the unfocused struggle of each against all for research funds. Existing centers needed to demonstrate productivity, or they might be closed down. Because the HGP was premised on achieving ongoing five- to tenfold increases in efficiency, a rate of production that was excellent one year might be totally inadequate the next.[24] In effect, each genome center was being compared not only with other centers but also with expectations, which blended seamlessly into hopes, about future productivity. Research groups who wanted to launch new centers found themselves competing with well-funded, existing centers, and they struggled to develop promising mapping strategies that would win a coveted center grant.

With the five-year goals emphasizing STSs, genome centers quickly began developing strategies for producing them, and in the United States the number of STS markers that a laboratory produced soon became a metric for measuring its output. At multiple organizational levels, genome scientists began using STSs to set goals and evaluate progress. At the annual Cold Spring Harbor meetings, genome centers started reporting the number of STSs generated. Centers also used STSs to monitor and motivate themselves. NATHAN, the director of EASTERN GENOME CENTER, made a game of challenging staff members to reach the laboratory's internal goals for STS production ("I bet you can't generate 1,001 markers by Cold Spring Harbor"), which exceeded the more conservative goals the center had promised its funders.[25] Funding agencies also used STSs to monitor progress; they provided, as TOM, the director of NORTHERN GENOME CENTER enthusiastically put it, a way to "hold people's feet to the fire."[26] Science policy makers also used STSs to assess overall progress in genome mapping. When the NCHGR published a report on the HGP called the "Genome Report Card" in 1992, it listed the numbers of STSs mapped to each chromosome along with several other metrics of progress.[27]

This is not to say that STSs completely eliminated the complexities of measuring progress in mapping.[28] In particular, counting numbers of STSs did not provide a measure of how close a physical map was to "completion." The issue of creating a quantified measure of the "doneness" of a map was tricky. As in conventional cartography, one genome map might cover a larger area, but another might be much more detailed. Which is the better map? Another issue was the question of whether knowing the general location of a mapped area is important to a measure of "doneness." When mapping land, for example, one might be able to create detailed maps of random parcels without knowing much about their general location. (Think of having thousands of detailed little maps made from aerial photographs of small bits of land randomly scattered across a continent; theoretically, once enough pictures are taken, they will overlap to create a complete map, but along the way one would have lots of disconnected fragmentary maps.) If a number of random regions have been mapped in this disconnected sense, how should that count in a measure of overall "doneness"?

Such questions had practical consequences because some genome centers were using mapping methods that tended to produce many highly detailed fragments of maps without shedding light on the location of those fragments on the chromosome. Critics argued that capturing 80 percent of a chromosome in a collection of unordered fragmentary maps did not mean that the map was 80 percent done. Various schemes for quantifying the distance from "completion" were proposed, and a metric focused on long-range continuity as the key measure of "doneness" ultimately won the approval of the most influential genome project advisers.[29]

Resistance

The STS standard was not uniformly welcomed, and some genome scientists, including a few prominent Europeans, offered criticisms.[30] One worry concerned how well STSs would work in practice. When the STS proposal was conceived in the summer of 1989, PCR remained a rather new technology.[31] PCR assays broadly similar to those underlying STSs were rapidly moving into widespread use, but the five-year plan admitted that "few, if any, mapping projects have started to use the STS system," noting that the STS proposal was "still under discussion in the scientific community."[32] Many techniques in molecular biology involve tacit knowledge and are sensitive to small changes in procedures or conditions, raising the possibility that the PCR assays underlying the STS system would work inconsistently in different laboratories. At the time the US genome program adopted the STS standard, for example, no one had actually demonstrated the feasibility of transporting STS markers from one laboratory to another merely by sending written information, without face-to-face interaction at the bench or a telephone call.[33]

Moreover, the STS standard imposed costs on genome-mapping laboratories. Before the standard was instituted, a laboratory that had put a clone-based marker "on the map" considered the job of positioning that marker to be all done; under the standard, most laboratories would now have to perform the additional step of making an STS based on that clone.[34] Although technically straightforward for a well-equipped laboratory, this step required sequencing a small piece of DNA from the clone. Some of my interlocutors who were building genome maps using clone-based strategies objected to the extra and unexpected costs of translating their markers into STSs. In addition, some gene hunters—an important set of users of HGP results—were unhappy with the prospect of having to perform PCR assays in order to use STS maps, in part because PCR techniques were new and unfamiliar.

Beyond cost, a few scientists also objected to the policy's managerial vision. In 1992, one well-known EUROPEAN GENOME SCIENTIST, in a rhetorical flourish drawing on the recent collapse of communism, compared the focus on quantified goals and production metrics to the failed command economies of eastern Europe.[35] To the architects of the STS proposal, however, the advantages of maps describable entirely as written inscriptions massively overshadowed the inconveniences of translating clone-based markers into STS form. Moreover, they saw quantified goals and uniform metrics of production as wholly beneficial, not undesirable. Proponents of the STS standard were willing to weather a bit of what some called "community resistance" and expected it to be short-lived.

In the United States, resistance soon vanished, not surprisingly in light of funding-agency policy. As PCR technology spread, quickly becoming an indispensable tool in molecular biology, map users' concerns dissipated. A high degree of transportability between laboratories was achieved.[36] Moreover, small laboratories liked being able to use maps without having to await the arrival of clones. Ultimately, making STS-based maps available to any and all—with no strings attached—played a key role in persuading critics of the HGP that its concerted mapping program had been worthwhile. In 1992, Nathan of Eastern Genome Center, described how his laboratory had won the hearts and minds of a skeptical community:

I think people were royally pissed off when we got the genome center grant. ... We knew there was a tremendous amount of resentment about that. ... I very concretely recognized that the only way to deal with the problem was through deed rather than word. So we got to LOCATION OF SCIENTIFIC MEETING, and we just handed out the map. That settled all the problems. A number of people came over and said, "That was a real coup." That settled all the resentments right there because whatever we had, we were sharing.[37]

Nathan, like many of the other genome center directors, wanted to compete not by hoarding clones and keeping maps secret, but by producing and distributing more genomic data faster than his competitors. Beyond distributing maps at meetings and publishing papers, genome centers began posting unpublished preliminary versions of STS-based maps on the Internet, an innovative practice at the time. The Whitehead/MIT Center for Genome Research (CGR) created an Internet-accessible database, which it used to distribute updates of its genome maps. To avoid the need to update the database constantly and to relieve users of the burden of constantly checking to see if new data had been added, this center developed a policy of releasing new data monthly: "At the end of each month, all genomic mapping data are reviewed and prepared for distribution via CGR's electronic databases. … CGR's data release policy aims to ensure that scientific colleagues have immediate access to information that may assist them in the search for genes."[38]

CGR's data-release policy also distinguished between scientific publication and appearance in the database, stating clearly that the center expected to prepare journal articles based on the data and to underscore that the data were provisional and might contain errors: "Data releases," the policy explained, "do not constitute scientific publication of CGR's work, but rather provide scientists with a regular look into our lab notebooks."[39] In fact, the maps distributed this way did not neatly fit into a conventional binary distinction between the published and the unpublished. On the one hand, the data were available to any and all, so in that sense they were published. On the other hand, the data had not appeared in peer-reviewed journals, and they were not embedded in the control relationships of the scientific publication regime. Moreover, CGR asserted that the data were unpublished, said it planned to publish them later, and endeavored to set limits on their use that reached beyond the minimal restrictions applicable to information published in journals. For example, CGR asserted that some uses, such as projects central to its own genome mapping research, could be undertaken only via a collaboration.[40]

The data that CGR posted arguably had a slightly ambiguous status in the peer-review process.[41] The center had published peer-reviewed articles on its procedures for generating data, and it had quality-assurance procedures (such as mathematical detection of potentially erroneous data points) built directly into its data-production system. It also conducted its own internal review prior to each scheduled release, with a signoff from the center director and several other scientists. But much of the data it released had not been formally peer reviewed by a journal. As this example suggests, new modes of data production and distribution can become occasions for constituting knowledge-control regimes governed by control relationships that do not neatly fit the categories of previously existing regimes, such as the one governing scientific publication.

Aligning with Molecular Biology

Governing the HGP entailed aligning the new regime closely enough with extant practices in molecular biology to make building this form of biology feasible. The factory-style laboratories that genome project leaders envisioned could not be instantly formed by scaling up the material practices of what my interlocutors sometimes called "cottage industry" biology. A path leading from the existing to the imagined had to be constructed. Joan H. Fujimura has described how constructing do-able problems in molecular biology entails aligning experiments with opportunities and constraints operating at other organizational levels.[42] Genome centers faced analogous problems. They could not depart too suddenly and dramatically from the prevailing culture and practices of molecular biology. For example, the early genome centers had to accept the constraints of their institutional environments. Asked about his philosophy for managing a genome center, Nathan emphasized the locality of the problem: "I don't know how to run a genome center. Nobody does—but, more importantly, the question for me is how to run a genome center *here*."[43] Local constraints varied. In some instances, organizational rules governing hiring and academic appointments shaped the makeup of the workforce. Laboratory space was also a problem. The first NIH genome centers had to fit into architectural spaces designed for ordinary molecular biology; when the most successful centers expanded after a few years, they moved into much bigger off-campus spaces, capable of accommodating dozens of scientists and laboratory personnel.

Genome centers also had to begin scaling up by building on the existing bench-top technology and social organization of "ordinary" molecular biology; they could not instantly implement the elaborate industrial-scale data-production "pipelines" that ultimately became emblematic of the field.[44] Genome centers introduced a variety of novel material practices intended to produce mapping data in a robust and rapid way. These practices included using easily implementable, small-scale automation to save time and reduce error; creating highly standardized protocols to move toward what some called "assembly-line" production; and, most importantly, interweaving the laboratory's work with an informatics system to track the flow of materials and data.

Creating genome centers also entailed making them fit with the goals and career expectations of existing pools of scientific talent. Because reaching HGP goals would require continuing technological improvement, genome research was not a matter of creating a production system and simply hiring enough technicians to operate it until the job was done; the task required ongoing technology development, involving a mixture of biological science, software design, mathematics and algorithm development, and systems engineering. Genome centers thus needed to attract and retain strong scientific talent, especially postdoctoral associates, graduate students, and junior staff scientists who could do the creative work of continually building and evaluating new systems. But at the outset of the HGP, career paths in genome research were not well established. The centerpiece of training in molecular biology was a postdoctoral or dissertation project that a young scientist could complete within the career-friendly timescale of just two or three years. Junior scientists also worried that working on the HGP might not yield authorship of the highly visible papers they needed for a successful scientific career. Genome center directors told me that they had to organize the work so that stellar young scientists—"who don't want to be somebody's peon"—would want to do it. Such people, Nathan explained, "can't do 10 percent of a map or something. They have to do 100 percent of something."[45] Long-term mapping projects would take too long and involve too many contributors to help early career researchers build a distinctive scientific identity. Thus, it was essential to organize work so that a "single individual" could make a "spectacular contribution" in about two years. For example, "SAM ... came to my lab as a graduate student," Nathan explained, "[and] put in probably two years gearing up and building the map. ... Sam was willing to do that because at the end of 16, 18 months he now has ... a tool that's spectacularly valuable to a lot of people. ... That's allowed him to set up collaborations with labs studying modifier genes for cancer, modifier genes for adult onset diabetes. ... Sam realizes that, as they say in the movies, 'if you build it, they will come.'"[46]

Becoming Durable

So far I have described how in the early years of the US project, new kinds of laboratories, new research agendas, and a new regime of governance were coproduced. The governing frame of a regime to guide the US human genome program effort took shape though a process that involved constituting a space in the science policy discourse, instituting the NIH and DOE programs, developing overall goals, outlining a five-year plan, writing and revising the genome center RFA, and requiring the use of STS landmarks to describe maps. The governing frame defined the "genome project" as distinct from "ordinary biology" and established the principle that these two types of biology were to operate under different systems of governance. It specified a set of key agents, objects, and actions and aimed to bind them together with a relatively tight set of control relationships. The agents included genome centers (and applicants who sought major grants to operate them); the funding agencies (that is, the NIH, along with its junior partner, the DOE genome office); the agencies' network of outside science advisers; the small laboratories of ordinary molecular biology; and the scientific community writ large. Such entities as project goals, grants and grant proposals, mapping data, and even specific mapping landmarks were among the more important knowledge objects. Salient actions included establishing and refining goals, producing map data, making maps available to users, reporting to funders and program advisers, and evaluating the output of genome centers and the progress of the program as a whole.

Although most of these agents, objects, and actions had analogs in the world of "ordinary biology" and in the regime of uncoordinated, investigator-initiated grants, the HGP substantially reframed the entitlements and burdens operative in genome research. It configured the funding agencies and their science advisers as active managers of the program, empowering them to discipline genome centers with accountability mechanisms unheard of in the small-laboratory culture of molecular biology. They would issue five-year plans, set specific goals, establish quality and production standards, evaluate progress continually, and adjust the research program as needed. The funding agencies' managerial rights imposed correlative duties on genome centers, which ranged from providing more frequent updates on progress than was typical in ordinary molecular biology to producing data according to specified standards. The scientific community as a whole was granted unencumbered rights to use the data that centers produced. In short, the knowledge-control regime for governing the HGP aimed to impose strong accountability requirements on genome centers without disturbing the regime governing ordinary biology, which was framed as a different jurisdictional space.

This regime quickly grew durable, especially as concrete demonstrations of the HGP's commitment to rapid data distribution shored up support for the project among biologists in many fields. As the focus of the HGP shifted to sequencing in the latter half of the 1990s, the key elements of its constitutional framework continued to operate. To be sure, there were changes. The boundaries of the HGP required continual policing and some adjustment, and new goals were articulated in 1993 and 1998 in response to ongoing developments. The NIH elevated the status of NCHGR in 1997, creating the National Human Genome Research Institute (NHGRI). In the late 1990s, there was a winnowing out of genome centers and a concentration of resources. But, despite change, the project continued to be governed by control relationships intended to keep the project on track and make genome centers accountable. A leadership composed of funding-agency officials, outside science advisers, and genome centers coordinated the project, keeping goals focused and, to the extent possible, quantitatively measurable. Producing data and technology remained genome centers' raison d'être, and rapid data release remained a persistent priority. Indeed, the US funding agencies intensified their role in coordinating genome centers during the final years of the project.

Building the Reference Library System

As governments and private charities outside the United States initiated genome programs at the close of the 1980s, each configured its program differently. The French organization Généthon, the United Kingdom's

HGMP, the European yeast-sequencing program, and the ICRF's Reference Library System in London all pursued somewhat different research goals, and the knowledge-control regime that each constructed differed from the others.[47] We will concern ourselves here only with the RLS because the knowledge-control regime it implemented offers an especially illuminating comparative case.

The architects of the RLS sought to move beyond the existing practices in molecular genetics by instantiating what they viewed as a more cooperative, collaborative mode of conducting genome research. This faction of the genomics vanguard aimed to constitute a knowledge-control regime that would speed research and reduce duplication of effort. The goal was to establish a network of laboratories bound together by a common set of biomaterials and a system of mutually beneficial control relationships. Let us examine the effort to constitute this new regime and the materials, technologies, procedures, and control relationships implicated in its governing frame.

Orderly Libraries

What evolved into the vision for the RLS began in Hans Lehrach's laboratory in the late 1980s. Lehrach sought to develop technology for conducting highly "parallel" experiments, designed to yield massive amounts of genome-mapping information in a single experiment. Lehrach, a critic of the individualistic culture of molecular biology and human genetics, believed that unraveling the complexity of biological systems would ultimately require more collective modes of organization. Duplication of effort was untenable given the challenges of mapping and sequencing and the urgency of identifying disease genes. He also thought that existing arrangements created incentives for scientists not only to move quickly themselves but also to slow down other scientists by hoarding data or renaming clones to obscure their identity. As Lehrach and his colleagues achieved success with their novel mapping technology, they began envisioning ways to use it to orchestrate new forms of translaboratory collaboration, a process that inspired the knowledge-control regime underlying the RLS. To understand the regime, then, it is necessary to take a close look at the technology.

Lehrach's laboratory constructed systems for manipulating libraries of clones with robots and for cataloging information about them in computerized laboratory databases. The laboratory stored libraries in microtiter plates—rectangular plastic plates about 3.5 inches by 5 inches, with 96 indentations, or "wells," arranged in 8 rows and 12 columns. Clones stored in plates became "addressable"; each could be unambiguously identified by row and column location as well as by a number designating the plate in which it was stored. The address doubled as a unique clone name for use in a computerized database containing information about the clone's characteristics. Robots used these addresses to transfer materials in and out of specific wells. Lehrach thus built an integrated system that facilitated storage and retrieval of clones. At the ICRF, the RLS set up a number of addressable libraries, beginning with cosmid libraries and later adding YAC and other clone libraries.

Lehrach's laboratory extended the familiar molecular biology technique of hybridization using labeled probes (recall figure 2.3) to create a new method, high-density hybridization, for quickly testing an entire library to find all the clones that contained sequences matching particular DNA "probes." These hybridization experiments used robots to position clones in tiny dots on flat sheets of DNA-binding material known as "filters." The robot could place about 9,600 samples in precisely positioned locations on a single filter (a very high density at the time). Next, a radioactively labeled probe would be poured onto the filter. The probe would stick (or "hybridize") to matching sequences on the filter, thus binding to the locations where the probe matched a clone. Placing the filter on top of a piece of X-ray film produced an "autoradiogram"—an image with dark spots indicating the locations of clones that matched the probe (figure 4.2).

The RLS deployed high-density hybridization in two main kinds of experiments. The first aimed to identify clones that contained particular probes—for example, a probe believed to lie near a disease gene. Such

---

Figure 4.2
Detection of clones using high-density hybridization. (**a**) A robot moves clones stored in microtiter plates, positioning them in precise locations on filters. (**b**) A radioactively labeled probe is poured over the filter; it hybridizes with clones containing matching DNA. X-ray film detects the positions where probes hybridized to clones. (**c**) The coordinates of spots that "light up" on the filter are used to identify the clones. A sample of the clone is transferred into a tube. (**d**) The clone is retested against the probe to eliminate any false positives. Illustration by Chris Cooley, Cooley Creative, LLC.

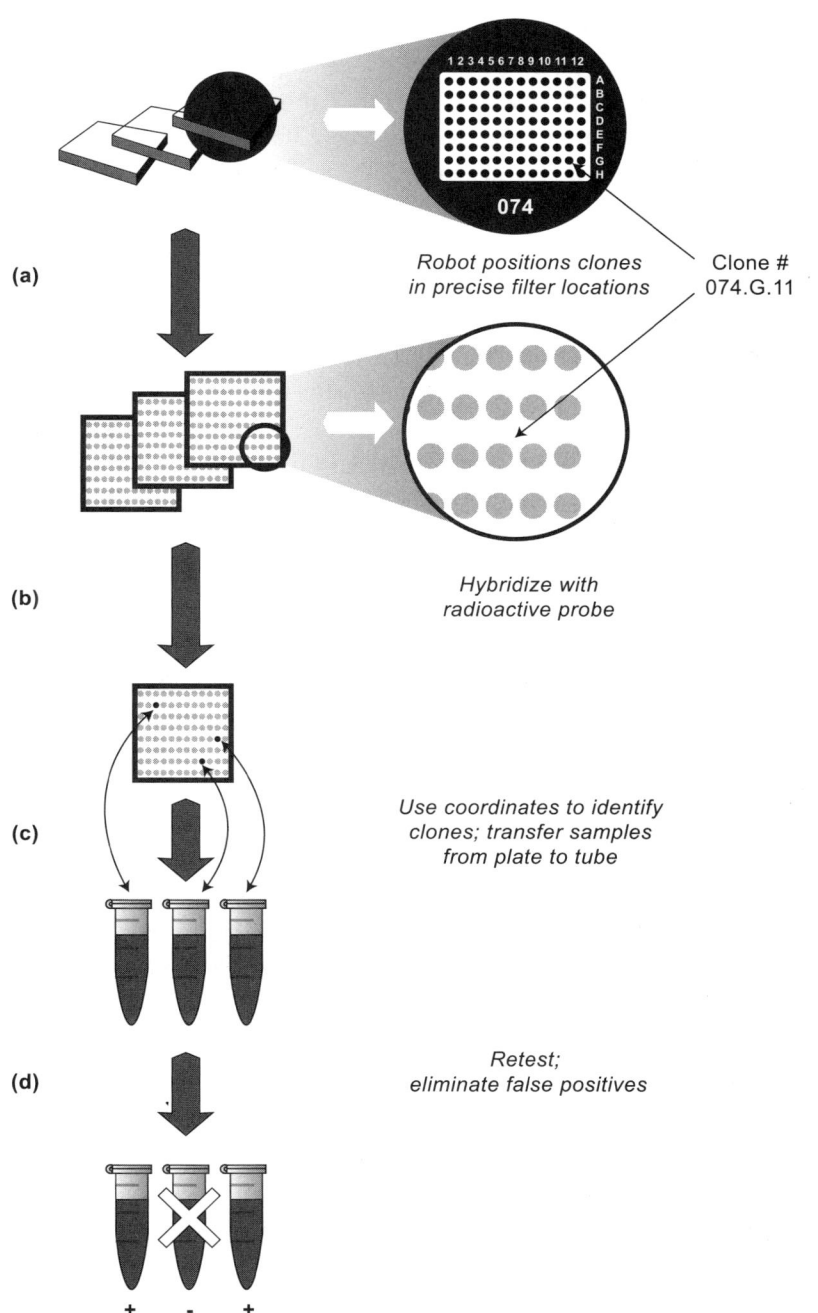

(a)   *Robot positions clones*   Clone #
           in precise filter locations   074.G.11

(b)   *Hybridize with*
           radioactive probe

(c)   *Use coordinates to identify*
           clones; transfer samples
           from plate to tube

(d)   *Retest;*
           eliminate false positives

           +   −   +

experiments could quickly yield a set of clones of great value to a gene hunter. The second type aimed to identify overlapping clones by revealing clones that shared DNA sequences. After enough information on overlaps had accumulated, computerized analysis could construct a contiguous array of clones. Lehrach and his colleagues demonstrated that this mapping strategy could work, at least with the relatively simple genomes of microorganisms, by creating a contig map of the yeast *Schizosaccharomyces pombe*.[48]

Orderly Translaboratory Relations

For Lehrach and his colleagues, the high-density array technology came to inspire not only biological experiments but also a social one. What if many laboratories did hybridization experiments with the same libraries and sent the information back to a central database? Data about these "reference" libraries would accumulate quickly. Scientists would pool useful but unpublished results. Duplication of effort would be reduced. Progress would accelerate. These ideas sparked the vision for the RLS, which aimed to choreograph exchanges among a community of laboratories, all of which would use the same libraries and filters. By mid-1989, the same year that the US program opted for the STS standard, the RLS had taken shape.

The RLS constituted a knowledge-control regime that integrated high-density hybridization with a system of rules intended to orchestrate productive interactions among two main categories of agents: the "central laboratory" and the "outside laboratories." Its governing frame cast Lehrach's laboratory as the central laboratory standing at the center of a network. Its role was to house and manage the libraries, maintain a central database, and provide materials and data to the outside laboratories, which the regime framed as the peripheral nodes of the network. Any laboratory could join the network in the role of a peripheral node. These outside laboratories (a.k.a. the "participating laboratories") were typically molecular genetics laboratories engaged in disease-gene hunting, not large centers of genome research. The knowledge-control regime configured these outside laboratories as both "users" of the RLS and "contributors" to it.

To choreograph interactions among these agents, the regime defined several types of routine transactions, each structured by a different script. In the first script (figure 4.3), the RLS distributes high-density filters to an outside laboratory, which performs hybridization experiments using its own

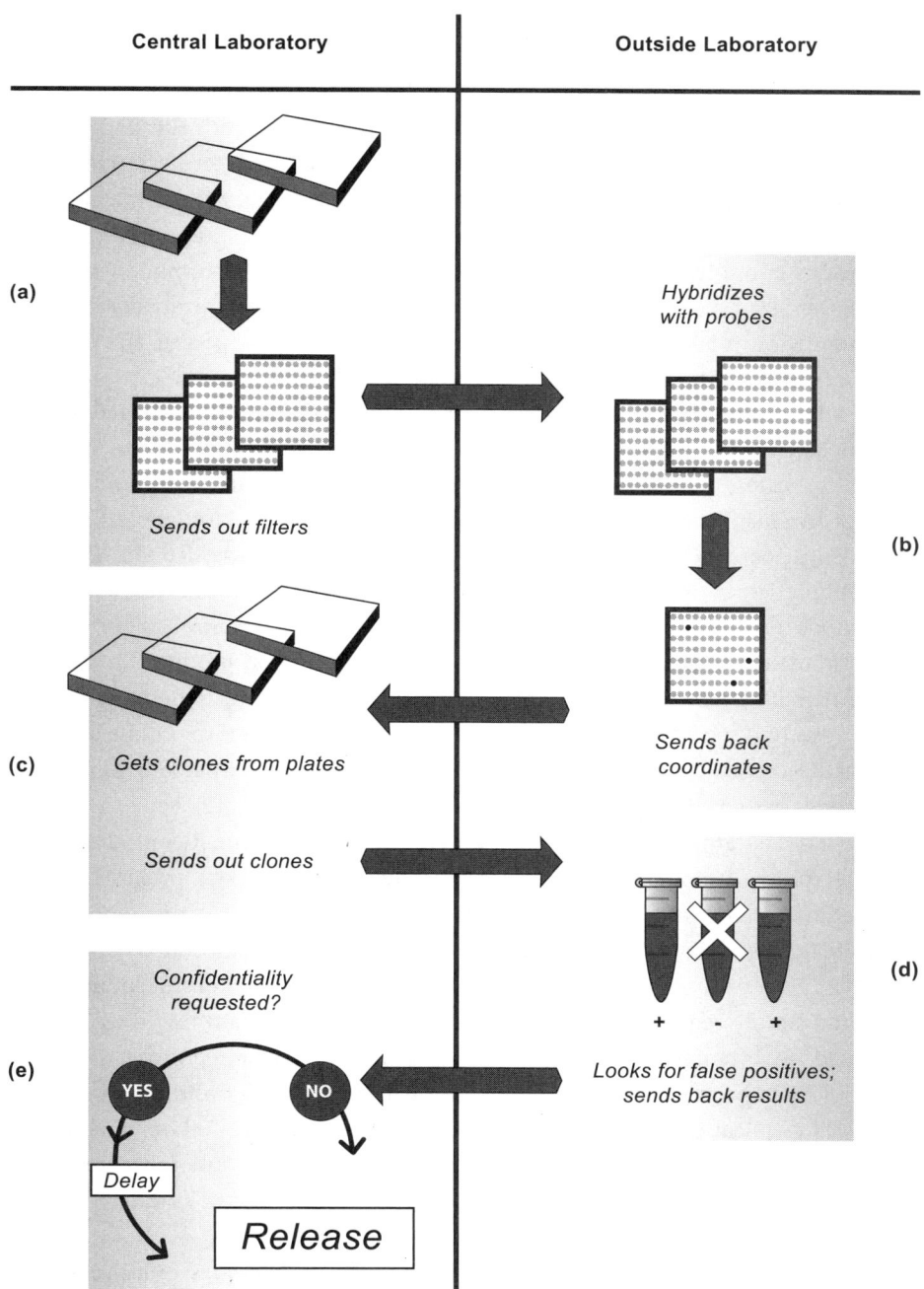

Figure 4.3
Script for an RLS transaction. Illustration by Chris Cooley, Cooley Creative, LLC.

probes (e.g., a probe believed to be near a disease gene). The outside laboratory then reports the results to the central laboratory by sending back the grid coordinates of the "positive signals." Using the coordinates, the central laboratory then identifies the corresponding clones and sends them to the outside laboratory. At this point, the outside laboratory tests the clone against the probe to confirm the result of the original experiment, a step considered necessary to protect against the "false positives" that occur regularly in hybridization experiments. The findings of the confirmation step are then sent back to the central laboratory, stored in the central database, and made available to all of the outside laboratories—perhaps after a period of confidentiality (discussed more fully later).

For the outside laboratory, transactions following this script offered an efficient way to find strategically valuable clones. In the early 1990s, the ability to test a probe against thousands of clones in a single experiment was awe inspiring, and this capacity provided an incentive for laboratories to participate in the regime. Through such transactions, the RLS aimed to help participating laboratories. Beyond this "service" function, these transactions were intended to benefit the entire RLS network—central laboratory and outside laboratories alike—by gathering information produced in the participating laboratories and making it available to the entire network.

A second type of transaction enabled outside laboratories to use the central RLS database to find clones relevant to their work. As participating laboratories tested filters with their probes, the information they sent to the central laboratory accumulated. Using this information, the RLS produced directories of clones. One directory indexed clones by the probes that matched them, and another used probes whose locations in the genome were both known and "available" to catalog clones by chromosomal region. Thus, a scientist seeking clones from, say, the q22 region of chromosome 21 could simply request them. Gene hunters thus could obtain useful clones based on other people's experiments—even if no paper reporting those results had been published in the scientific literature, thus avoiding unnecessary duplication of results. The RLS sought to produce collective benefits for the network as a whole and at the same time to contribute to the individual research projects of each outside laboratory. Moreover, as data accumulated, the utility of the reference libraries would increase.

The RLS also facilitated transactions of a third form: requests for probes from one outside laboratory to another. Gene hunters often restricted access to their probes—especially their most precious ones—and the RLS did not attempt to collect the probes. However, the central database included a list of laboratories that possessed probes from different regions of the genome. Consulting the "List of Probes Ordered by Chromosome Location" enabled the user to see, for example, that E. M. VAN GALDER had four probes in the q22 region of chromosome 21. In this way, researchers could learn of probes relevant to their work and find out whom to contact about their "availability."

Although the RLS knowledge-control regime aimed to draw some data together for the entire network to use, it also established boundaries that allocated control over other knowledge objects. Entire libraries, the core resource of the central laboratory, were distributed, if at all, only to the closest of collaborators.[49] The central laboratory's control over the libraries was backed by the system's material properties: filters containing DNA from an entire library could be widely distributing without transferring control of the library because it was impossible to replicate the library from the filters.

The system's material features also reinforced the control relationships governing RLS transactions. Once an outside laboratory elected to participate, received filters, and conducted hybridization experiments, it could obtain useful information *only* by sending the results of its experiments back to the central laboratory. The autoradiograms displayed nothing more than meaningless dots until the central laboratory entered their coordinates into the database, which linked filter positions to the underlying clones. When participating laboratories sent their hybridization results back, they also provided the central laboratory with information that enhanced the libraries' utility. In this way, the RLS built accountability and control directly into its material structure.

The RLS also explicitly defined the control relationships pertaining to the experimental results produced using its materials and information. For example, the central laboratory made it emphatically clear that "the distribution of filters and clones from the Reference Library does NOT establish a collaboration (e.g. co-authorship)."[50] The central laboratory thus stressed that it did not seek to intrude into the autonomous space of the outside laboratory with demands of authorship. Outside laboratories were asked merely to send reprints of publications arising from the use of RLS materials and to acknowledge the origin of any clones used—both normal courtesies in molecular biology. In a reciprocal way, the statement also denied to outside laboratories any claims of coauthorship on RLS publications.

Because gene hunters were possessive of their "unique," "private" probes, the RLS offered participating laboratories the option of designating some results "confidential," promising that no information about a confidential probe or the clones it "hit" would be disclosed for six months (expandable to a year and even beyond in some instances). However, the RLS would register the existence of confidential information in the database and make public the fact of its existence.[51] Until then, no further details would be released. Many outside laboratories made use of these confidentiality provisions. In early 1991, some 24 percent of the results listed in the RLS database provided incomplete information owing to confidentiality.[52]

A Road Not Taken

When the leadership of the US genome project was considering how to structure its mapping program and which technological strategies to pursue, many options were possible. The RLS approach was one of them, but the US genome decided to opt for the STS approach. To some extent, the question was a technological one.[53] For example, some genome leaders argued that hybridization had an unfavorable signal-to-noise ratio, making the high-density arrays less "robust" than the PCR-based STSs.[54] But the normative dimensions of the question should not be neglected. Members of the US genome leadership objected to depending on a central laboratory that planned to house "everyone's data" centrally while promising to provide the data on request. Over the next few years, these worries persisted. As Lew, the director of Coastal Genome Center, put it in 1993, even if the central laboratory does make the data available, "I don't want to rely on one place to do this. I don't think it's the way it should be done."[55] Some critics raised questions about the integrity of the system: What if confidential information gets leaked? How do we know that the databases will remain secure? Could the central laboratory suddenly change the rules? To many researchers in human genetics and genomics, a mapping strategy that required this degree of dependence on—and trust in—a central laboratory seemed naive.

The RLS was, however, able to attract many users.[56] By early 1991, about 100 laboratories had received clones from the central laboratory, and soon

more than 400 scientists had participated in the system. The RLS ultimately distributed tens of thousands of clones—more than 46,000 by the
end of 1993—and even critics, such as Lew, acknowledged that the RLS
had "got a lot of stuff out to people" and had a real impact on the field.[57]
However, the system never functioned so well in execution as envisioned
in conception, and participating scientists acknowledged that the system
was not running completely smoothly.[58] For one thing, some outside laboratories renamed probes and used the new names, not the publicly available ones, when reporting their findings to the RLS, thus making it hard to
build on their results. Many outside laboratories also proved slow to send
back information. In particular, after receiving the clones that their probes
"hit," outside laboratories often proved slow to return the results of the
confirmation step intended to eliminate false positives. In such cases, the
central laboratory could do little besides admonish users. In 1991, the RLS
newsletter noted that confirmations were missing on 86.8 percent of all
the clones sent out to date. "One of the main aims of the Reference Library
System is … to reduce unnecessary duplication of work. Therefore please
let us know the results as soon as you have analyzed the clones you have
received from us and inform us if they are real positives. … Some users are
good in this respect but others are slightly reluctant."[59] The newsletter
also included a list of all the scientists who had received clones and the
status of each clone—confirmed positive, false positive, and "potential
positive" (read, awaiting confirmation). "Should it occur that there are one
or two clones for which you have not sent the analysis results to us, *please
do it as soon as possible.*"[60] Of course, no one could tell which laboratories
had yet to perform the tests and which were "slightly reluctant" to provide
results.

Most of the outside laboratories that participated in the RLS appear to
have been motivated by a desire to obtain some strategically valuable clones
rather than by a commitment to the RLS's broader vision of translaboratory
cooperation. In 1994, Lehrach left the ICRF to become director of the Max-
Planck Institute for Molecular Genetics in Berlin, where he continued to
promote RLS-style science. In 2001, a scientist intimately familiar with the
history of the RLS argued that the original vision "was scientifically sound
but didn't take account of the personalities of the average scientist, who is
maybe a little bit less generous and altruistic than expected and maybe a lot
more paranoid."[61] Although many clones were distributed, the RLS failed to

constitute a durable knowledge-control regime, and the broader changes that its architects hoped to enact did not materialize.

Conclusion

The sociotechnical vanguards that constitute new knowledge-control regimes build normative visions of orderly knowledge and orderly social relations into sociotechnical structure, and comparative analysis is a useful tool for appreciating how this is achieved. Comparing the knowledge-control regimes of the RLS and the HGP is revealing in part because of the many similarities in how they approached problems of control and coordination in the production and use of genome maps. Both regimes emerged at the same time, and both aimed to establish a set of control relationships that would support orderly interactions between (one or more) genome centers and the small laboratories typical of molecular biology. To achieve this, both regimes sought to constitute agents with specific burdens and entitlements, interweaving rules with particular genome-mapping technologies in ways that made the material structure of the system into part of the means of enforcing its rules. But despite these similarities, each regime took a distinct approach to choreographing the translaboratory flow of knowledge objects, and each built into its sociotechnical structure a very different normative vision of the proper relationship between centers of genome research and other laboratories.

The most important differences between the US genome program and the RLS are highlighted in table 4.1. The US genome program constructed a regime designed to hold genome centers accountable—using productivity metrics and data reporting and release requirements unusual in molecular biology at the time. Meanwhile, the regime left virtually untouched the structural position of small laboratories vis-à-vis large centers: small laboratories could draw on STS-based genome maps freely without incurring obligations or becoming entangled in negotiations with centers. At a time when the HGP remained controversial, this regime offered a means to reassure those who feared that genome centers would squander funds, hoard data, or use their abundant resources to dominate smaller laboratories.

But if the HGP regime aimed to establish strong control over genome centers without imposing controls on small laboratories, the RLS aimed to achieve almost the reverse. The RLS vision of a less-individualistic genomics entailed drawing participating laboratories into a network bound together

Table 4.1

Comparison of the HGP and RLS Regimes

US Genome Program	Reference Library System
Strong institutional backing; large budget.	Weak institutional backing; small budget.
Genome centers are configured as data producers, outside laboratories as users.	Central laboratory and participating laboratories are configured as components in producer/user network.
One-way flow of data: from center to "everyone."	Two-way flow of data and materials: from center to outside labs and back.
Outside laboratories need not interact with center; no long-term interdependencies.	Ongoing interactions link center to participating labs; center/periphery structure remains durable.
Seeks to control genome centers and sustain traditional molecular biology community practices.	Seeks to reform molecular biology community, making genomics more collective than molecular genetics.
Outside laboratories seen as an unorganized population of autonomous agents.	Outside laboratories seen as participants in an ongoing collective.
Employs technology to free users from needing connections to the center or to each other.	Employs technology to bind users to the center and to each other.

by new control relationships and an ongoing and bidirectional flow of materials and data. Whereas the US regime constituted genome centers as data producers whose raison d'être was to provide unencumbered resources to outside laboratories to use for their own particular purposes, the RLS imagined its central laboratory as a vital hub that integrated individual laboratories into a more collective knowledge-producing network. Whereas the US regime sought to speed the one-way flow of data to outside laboratories, the RLS sought to compel outside laboratories to return data to the central laboratory, the central node of the community. Whereas the US regime employed a PCR-based technology as a means to compel genome centers to export mapping data to the outside world, the RLS employed a hybridization-based technology as a means to require participating laboratories to send mapping results to the central laboratory. The regimes thus sought to build radically different orders for controlling relations between genome centers and the scientific communities that they were intended to serve.

An important reason for the stark differences in these social orders is that the two regimes expressed contrasting ways of imagining the scientific community.[62] It would not be much of an exaggeration to say that the US regime was premised on an imaginary of the scientific community as a *population* of autonomous laboratories, disconnected agents interested in maximizing their individual freedom of action. Similarly, without greatly overstating the case, we might say that the RLS was premised on an imaginary of the scientific community—at least potentially—as a *group* of laboratories bound together by common goals, technologies, and materials and interested in maximizing their collective achievements.

Although both of these ways of framing the scientific community are familiar features of scientists' discourse, the culture and practices of molecular biology lean heavily toward individual autonomy. The scientists who built the RLS thus envisioned an ambitious and broad transformation—one that aimed at nothing less than building a more collective research system that departed significantly from the prevailing orientation of molecular biology. That this effort failed largely reflects the fact that fully realizing this aspiration would have required significant shifts in the existing culture and practices of molecular biology. The US leadership, which could have chosen to go this route, actively rejected it on moral as well as technical grounds. Moreover, many of the outside laboratories that elected to participate in the RLS never fully bought into its vision.

In contrast to the RLS, the changes that the US program introduced at the outset of the HGP were consistent with preserving, not altering, the individualistic culture and practices of molecular biology. The US program constituted genomics as "different" from "ordinary biology," framing a regime with limited jurisdiction, making genome centers demonstrably accountable, and configuring them as sources of data and materials that small, autonomous laboratories could freely use without incurring obligations. The extant practices of molecular biology were deeply intertwined with institutionalized structures in US funding programs, universities, and scientific career paths. It is therefore likely that building a regime that did not challenge these practices—at least in the short term—was necessary to reach a settlement that would allow a US genome program to proceed and, in the longer run, to play a central role in significantly reordering the wider world of biological research.

5   Objects of Transformation

Transformative scientific change constitutes new knowledge objects and remakes old ones, contributing to the emergence of new ways of knowing and doing. Old paradigms are overturned, and the theories, instruments, objects, and objectives associated with them are radically reconfigured. But the laboratory is not the epicenter from which all such changes spread—although clearly it is a crucial participant in the process. Changes in policy paradigms, legal framings, and business models are also deeply implicated in creating new knowledge objects, reconfiguring old ones, and reordering the strategic objectives that these entities are used to pursue.[1]

This chapter examines an important set of knowledge objects that emerged as factions of the genomics vanguard took shape and contested the future of genome research, a process that constituted and reconfigured both knowledge objects and knowledge-control regimes. The objects this chapter examines involve information derived from sequencing complementary DNA (cDNA) clones—that is, clones made from DNA that codes for proteins (figure 5.1) and therefore is all or part of a "gene" (recall figure 2.2). As an example, consider the 308-base-pair sequence displayed in figure 5.2. This sequence is part of a cDNA; the full cDNA is much longer. Genome scientists might describe this sequence as a "partial cDNA sequence" (a.k.a. "partial cDNA") or as an "expressed sequence tag," (EST). Other significant cDNA-related objects that emerged during the HGP include "the cDNA strategy," "EST patents," and "EST databases." My account traces the emergence and transformation of these objects as well as the construction of regimes associated with them. During this process, factions of the genomics vanguard struggled over a variety of controversial questions: What *are* these objects? What uses and strategic value do they have? Into what sorts of knowledge-control regimes do they properly fit?

Who can use them to impose what forms of control over the future of genomics? As actors contested these questions, new configurations of objects, regimes, agents, and objectives arose, and genome research was significantly transformed. I focus my inquiry around six cDNA-related objects.

Object 1: The cDNA Strategy

Our first object—the cDNA strategy—emerged during the 1980s debate about what kind of human genome project, if any, to launch and how to structure and control its research agenda. Forged amid debate about the proper shape of knowledge-control regimes for governing national programs of genome research, the cDNA strategy was a hybrid scientific and policy object relevant to debates about feasibility, efficiency, priority setting, and program design. Advocates of limiting the size of the genome project to preserve research funds argued that it was unnecessary to sequence the entire genome. They contended that the information of greatest biological interest was in the "genes"—a term interpreted (in keeping with prevailing definition of a "gene") as including only DNA that coded for proteins.[2] In the United Kingdom, for example, Nobel laureate Sydney Brenner argued that the arduous task of sequencing continuous stretches of genomic DNA to obtain the "complete" sequence could not be justified absent faster, less-expensive technology. He called for sequencing only the protein-coding regions, which make up a small fraction of genome. "If something like 98% of the genome is junk," Brenner maintained, "then the best strategy would be to find the important 2%, and sequence it first."[3] Why spend billions of dollars sequencing "junk" if systematically sequencing cDNA clones would yield the sequence of all of the human genes, estimated at the time to number about 100,000? The "cDNA strategy" called for a genome project focused on sequencing "the genes" by sequencing cDNA clones.

The ensuing debate about whether to undertake "complete" "genomic sequencing" of the "whole genome" versus "cDNA sequencing" to acquire the sequence of "all of the genes" involved two conflicting ontologies. The cDNA strategy rested on an ontology that divided the human genome into "genes" versus "junk"—making those terms (and even the less-loaded terms *protein-coding region* and *noncoding region*) into policy concepts as well as

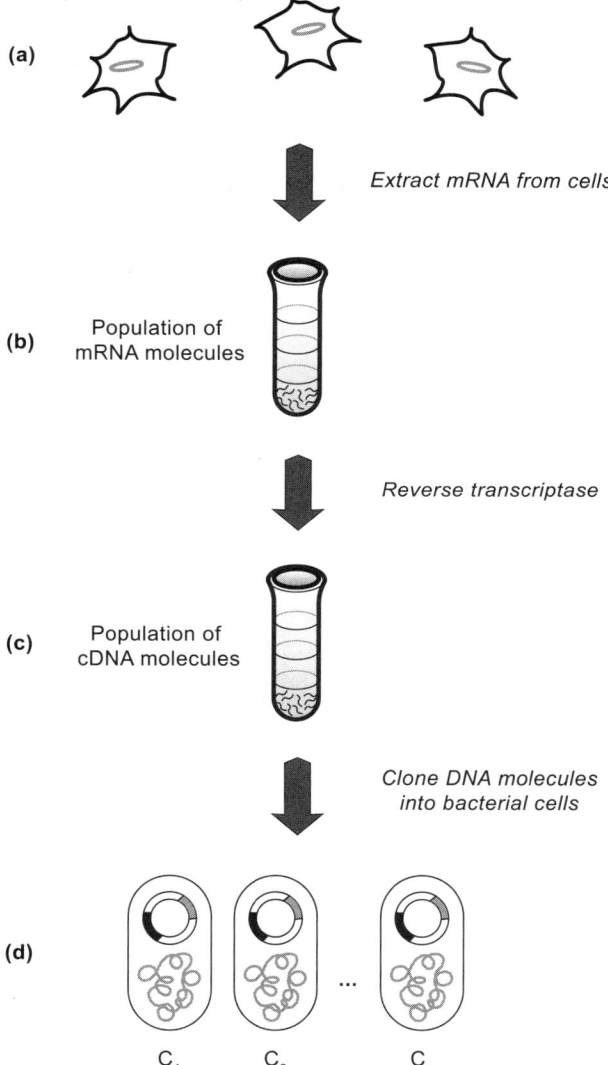

(a)

Extract mRNA from cells

(b) Population of mRNA molecules

Reverse transcriptase

(c) Population of cDNA molecules

Clone DNA molecules into bacterial cells

(d)

C_1 C_2 ... C_n

Figure 5.1

Creating a library of cDNA clones. Scientists extract total mRNA from cells **(a)** of the tissue type chosen (e.g., human brain cells). The resulting population of mRNA molecules **(b)** is treated with the enzyme *reverse transcriptase*, which forms DNA molecules with sequences that complement those mRNA molecules. The resulting cDNA molecules **(c)** are then cloned into bacterial cells **(d)** using recombinant DNA techniques (see figure 2.4). In this way, genome scientists can create libraries of clones containing DNA sequences expressed in the specific tissue type chosen. By beginning with mouse kidney cells, for example, a researcher can create a library of cDNAs made from genes expressed in the mouse kidney. Each cDNA clone may contain only a partial sequence of gene it encodes, not the full-length sequence found in the original genomic DNA.

```
AGAAGGCTCAGGATGAGATCCCAGCACT
GTCCGTTTCCCGGCCCCNNACCGGCCTG
TCCTTCCTGGGCCCTGAGCCTGAGGACC
TGGAGGACCTGTACAGCCGCTACAAGAA
GCTGCAGCAAGAGCTGGAGTTCCTGGAG
GTGCAGGAGGAATACATCAAAGATGAGC
AAAAGAACCTGAAAAAGGAATTTTTCCA
TGCCCAGGAGGAGGTGAAGCGAATCCAA
AGCATCCCGCTGGTCATCGGACAATTTC
TGGAGGCTGTGGATCAGAATACAGCCAT
CGTGGGCTCTACCACAGGTTCCAACTAT
```

Figure 5.2
A partial cDNA sequence. This cDNA sequence is a fragment, 308 bases long, of a
gene expressed in the human uterus. It appears as the following Genbank entry:
"EST41904 Human Uterus Homo sapiens cDNA 5- end similar to human tat-binding
protein isolog YTA2, mRNA sequence." This entry is linked to a major paper by Craig
Venter's group: Adams, Kerlavage, Fleischmann, et al. 1995.

scientific ones. Advocates of sequencing the "entire genome," in contrast,
employed an ontology that framed a genome as an object that could not be
divided a priori into important and unimportant parts. It had to be understood as a whole.

In the United States, the NRC report published in 1988 rejected the
cDNA strategy on several grounds. First, no one knew with certainty that
noncoding regions of "no apparent significance" were really "junk"; they
might have vital but unknown biological functions. Second, a cDNA-
sequencing project faced technical hurdles. Existing cDNA libraries over-
represented genes coding for proteins that are abundant in the cell, so genes
for rare proteins might be missed. Moreover, because existing cDNA libraries included many clones containing only fragments of genes, sequencing
clones from those libraries would yield many partial cDNA sequences, and
one might have to sequence many of them to assemble the "full-length"
sequence of a gene. The NRC committee strongly recommended sequencing the entire genome, although it allowed that cDNA sequencing could be
a useful "interim" strategy while waiting for sequencing technology to

improve.[4] In contrast, the cDNA strategy was incorporated into the UK HGMP, not least owing to Brenner's influence.[5]

When the HGP was officially launched in 1990, the NCHGR leadership, strongly committed to keeping its research program on course, did not include cDNA sequencing in its five-year goals. However, critics continued to push for sequencing only the "genes."[6] Some supporters of sequencing the whole genome also argued that cDNA sequencing deserved a share of NIH's resources, although others worried that a large-scale cDNA-sequencing project might undermine support for completing the entire human sequence. Ambiguities complicated matters. One could read a call for an "interim" strategy of cDNA sequencing as a euphemism for deferring genomic sequencing indefinitely.

France and Japan, along with the United Kingdom, also undertook cDNA projects of one form or another, lending credibility to proposals to add a cDNA component to the US program. cDNA sequencing gained further ground when the DOE announced in 1990 that it would devote some funds to cDNA projects.[7] In this context, the NCHGR decided to reexamine the role that cDNAs might play in the NIH program, and in late 1990 it convened an ad hoc committee that included some advocates of supporting cDNA projects and some who believed that the NCHGR "should proceed cautiously." The committee report identified some areas of cDNA research worth pursuing but concluded that cDNA sequencing "should not be a major part of the NIH Human Genome Program." "The long-term goals of the HGP should not be compromised for short-term payoffs, especially since those long-term goals are difficult to achieve. Full-scale pursuit of cDNAs may result in too much dilution of effort just to get quick biological returns that will come anyway from the regular research programs supported by NIH," the committee decided. To sequence all the cDNAs, the committee noted, was a project equivalent to sequencing an entire human chromosome—an undertaking then regarded as too ambitious given extant technology.[8] The committee also noted techno-logical challenges, such as making better cDNA libraries, getting full-length cDNAs, and avoiding redundant sequencing of genes for abundant proteins.

The NCHGR decided to stay the course. This is not to say that its leaders saw no scientific value in cDNA sequencing, but in their strategic imagina-tion a major commitment to cDNA sequencing posed a threat to a knowledge-control regime designed to ensure disciplined pursuit of carefully bounded, long-term goals. Some scientists continued to advocate the cDNA strategy for several more years.

Object 2: Expressed Sequence Tags

Everyone involved in debate about the cDNA strategy, advocates and opponents alike, viewed partial cDNA sequences (like the one in figure 5.2) as incomplete "fragments," mere stepping stones on the path to obtaining the "full-length sequences" of genes. The goal of the cDNA strategy, after all, was to sequence all of the genes in their *entirety*, not just pieces of them. However, in 1990 and 1991 Craig Venter and his colleagues repackaged partial cDNA sequences as *tools* for "indexing"—and even "finding"—genes. They dubbed these tools "expressed sequence tags." I use the term *repackaged* not merely to make an analogy to marketing, a world in which a label can transform the product (although *expressed sequence tag* did turn out to be an excellent promotional term) but also to capture Joan H. Fujimura's usage, which emphasizes "packages" that assemble theory and methods—and, in my account, modes of control—into coherent means of making particular types of knowledge.[9]

In 1990, Venter was at the NIH's National Institute of Neurological Diseases and Stroke, where he ran a laboratory doing genomic sequencing of a small region of the human genome. Venter was also busy trying to obtain funding from the NCHGR and was growing increasingly frustrated with the slow pace of both genomic sequencing and the NIH bureaucracy. In his autobiography, Venter recounts how he suddenly realized that he had been sequencing "the wrong DNA."[10] Rather than struggling to assemble the ABI machine's short reads (approximately 400 base pairs long) into a continuous genomic sequence tens or hundreds of thousands of base pairs long (figure 2.6), he asked himself, what if he picked cDNA clones at random and ran each one through a sequencing machine just once, without trying to get full-length cDNA sequences? The machines would produce some 400 bases of sequence data from the end of each clone. Each of the resulting partial cDNA sequences would be far too short to cover an entire gene, but it would still contain valuable information—indeed, more than enough information to search GenBank for matching genes that had already been found.[11] Each database search would produce one of several outcomes: (1) a matching human sequence—a "hit"—in the databases indicating that the sequence was from a *known human gene*; (2) no matching human sequence, and in those cases one could conclude that the cDNA must be part of a *previously unknown human gene*; (3) a matching sequence from a nonhuman organism—an outcome that might provide *clues about the human gene* from which the cDNA had been derived. Knowing that a human cDNA closely matches a gene for particular protein in yeast, for example, might provide hints about the gene's function in human cells.

Dozens of cDNA clones could be run through an ABI sequencing machine each day, and partial cDNA sequences would quickly accumulate. Moreover, by selecting cDNAs produced from a specific tissue, such as human brain, Venter could focus on genes expressed in the brain and thus produce a kind of "index" of brain genes. In addition, with PCR, each cDNA sequence could also be used to test DNA samples for the presence of the gene that it tagged. Venter bought a commercially available cDNA library made from human brain tissue and began sequencing. His group was soon "indexing" 20 to 60 new genes each day.[12]

By early 1991, Venter's group began preparing a paper on the technique of indexing genes using ESTs—a name that emphasized not the fragmentary nature of partial cDNAs but their informational content and use. The paper appeared in *Science* in June 1991. It described how searching the databases for sequences matching ESTs from 600 cDNA clones had identified sequences representing 337 "new genes."[13] About 14 percent of these genes were similar to known genes in other organisms, which yielded clues about their function. Arguing that whole genome sequencing would not find the majority of human genes for at least a decade, the paper concluded that ESTs offered a "fast approach" that "will facilitate the tagging of most human genes in a few years at a fraction of the cost of complete genomic sequencing."[14]

Venter's claims about speed and cost recast the still incompletely resolved debate over the cDNA strategy in terms of new objects: EST sequencing versus genomic sequencing. A news article in *Science* reported: "Craig Venter says he can *find all the human genes* for a fraction of the cost of the Human Genome Project" (emphasis added).[15] Venter declared that his approach was "a bargain by comparison to the genome project," while also saying that he saw ESTs as an adjunct, not an alternative, to genomic sequencing. He admitted, though, that the EST approach would miss some genes, suggesting he might find 80 to 90 percent of them.[16]

In the ensuing debate, critics raised questions (familiar from the debate about the cDNA strategy) about both redundancy and the number of genes an EST sequencing strategy might miss. Some scientists also questioned the significance and novelty of ESTs. The regimes that award credit for scientific contributions, like all knowledge-control regimes, are subject to the vagaries of interpretive flexibility. Expert communities answer questions about the value of scientific contributions through negotiations that sometimes develop into heated disputes.[17] Venter's critics argued that "finding" an EST was not the same thing as "finding" a gene because an EST does not reveal either the gene's location or its full sequence. They also defined the process of generating ESTs as "routine" work requiring little creativity; some described the idea of searching databases for matches to partial cDNA sequences as obvious—an argument that acquired greater importance after debate began about whether ESTs were patentable. Venter's critics maintained that he had merely applied standard sequencing techniques to a commercially available cDNA library, produced short bits of sequence, and then queried publicly available databases using standard search techniques that anyone with an unknown DNA sequence would normally employ.[18]

In a sense, however, these criticisms missed the central point. What was most revolutionary about Venter's ESTs was the very fact that the underlying techniques *were* completely routine. In an inversion of the injunction to pick the right tool for the job, Venter had transformed the job to match the tool.[19] The ABI machines easily produced the short cDNA sequences needed to generate ESTs. In contrast to genomic sequencing, there was no need to struggle to merge small pieces into much longer strands. When positioned in the assemblage of readily available entities that Venter had constructed, the sequences coming off the machine *were* ESTs, immediately usable as such with almost no additional work. Accepting the limitations of the existing cDNA libraries enabled work to proceed at once, without investing effort to improve them. In short, by building the job around the tools at hand, Venter was able to produce *many* ESTs *quickly*. Put another way, ESTs made it possible—again in conjunction with standard bioinformatics techniques—to extract additional value from the public databases. To be sure, some scientists continued to question the innovativeness of

Venter's work. But by routinizing the production of partial cDNAs and repurposing them as ESTs, Venter effectively transformed them into a tool for "tagging" or "indexing" genes—which he discursively equated with "finding" or "discovering" them.

The speed with which a single well-equipped laboratory could "crank out" ESTs accelerated the move from the gene-level perspective of molecular genetics to the genome-level paradigm that was the raison d'être of genomics. Venter's public declarations (e.g., that ESTs offered a low-cost route to quickly obtaining the "most valuable" genomic information) infuriated some HGP scientists, who continued to worry about undermining support for sequencing the entire human genome. Venter publicly advocated obtaining the full sequence in due course, and he and his allies sometimes described the HGP leadership's failure to fund large cDNA projects as narrow-minded.[20] The standard against which ESTs should be measured, Venter and his colleagues insisted, was not the information that the complete genome sequence would eventually provide but rather the limited information available at the time.

Object 3: EST Patents

Our third type of object—EST patents—became the center of an intense and international controversy after the NIH filed for patents on Venter's ESTs. Since the end of the 1970s, the American sociotechnical imaginary of generating progress and wealth through innovation had been married to new policies aimed at increasing national economic competitiveness by capturing commercial advantages from research. During the 1980s, the United States implemented a variety of policy measures that strengthened protection of intellectual property, and the scope of patentable subject matter expanded (amid ongoing controversy) to include genetically engineered life forms and genes that had been isolated and cloned. The Bayh–Dole Act of 1980 had encouraged universities and other research institutions to file for patents on inventions that their scientists produced. And a growing number of universities and research-funding agencies had established technology-transfer offices dedicated to securing rights to and licensing inventions.[21] The concept of "EST patents"—initially as a hypothetical hybrid of the scientific and the legal that actors might be able to create—arose in this context. ESTs immediately became embroiled in the ontological politics so central to patent policy.[22]

The idea that ESTs might be patentable "inventions" was highly nonobvious to most genome scientists.[23] Accounts of the EST patent debate credit Reid Adler, the director of the NIH technology-transfer office, with instigating the EST patent applications—something Venter himself had not initially imagined.[24] Adler, who had learned of Venter's EST work prior to the appearance of the *Science* paper in June 1991, theorized that ESTs might well qualify as patentable under plausible interpretations of existing law. He further reasoned that the NIH might be legally obligated to apply for patents on them under existing technology-transfer policies. Finally, Adler theorized that publishing ESTs—and thereby placing them in the public domain—might transform finding the corresponding full-length genes into what patent law would classify as a trivial inventive step, rendering those genes unpatentable. He regarded this outcome as highly undesirable, in keeping with the prevailing industry view that gene patents are essential to developing products and expanding the US biotechnology industry.[25]

In May 1991, according to Robert Cook-Deegan's account, Adler requested a meeting with Venter, and shortly thereafter the NIH quietly filed for patents on several hundred ESTs.[26] Venter's *Science* paper was published in June, and in July Venter disclosed at a congressional meeting that the NIH had filed for patents. This announcement startled even well-informed observers, and the extent of the surprise underlines the fact that HGP scientists had not imagined partial cDNA sequences or ESTs as intellectual property.[27] At this point, "EST patents" remained theoretical because no patents had as yet been issued, but many actors began to respond to the possibility that they might be granted.

The reconceptualization of ESTs as potentially patentable was an important innovation that thoroughly transformed their significance. ESTs were no longer merely a promising tool for indexing genes but part of a vanguard vision of a novel technological-legal instrument with the potential to reallocate control over the human genome. Observers had expected that the HGP would raise controversial legal questions about intellectual property.[28] But given the slow pace of sequencing, few anticipated that patent-based monopolies might concentrate control over much of the genome so *soon*. For two reasons, the prospect of EST patents made this outcome seem plausible. First, it was clear that if the claims in the NIH application held up, patent holders would enjoy extensive control. Despite the fact that the patent application was based on fragments of cDNAs, the NIH claimed ownership not only of the ESTs themselves (the partial cDNA sequences) but also of the full-length cDNAs (the "genes") that the ESTs indexed as well as of the proteins for which those genes code and of related antibodies. Second, EST production was already fast and seemed poised to accelerate.[29] At a time when finding just one disease gene took years, the NIH was claiming what looked like an enormous number of genes: 377 in its first application and soon thereafter 2,375.[30] Back-of-the-envelope calculations suggested that a well-financed actor could purchase dozens of sequencing machines, hire and train technicians, and begin "cranking out" tens of thousands of ESTs per year.[31] If EST patents were allowed, then many—perhaps even most—of the estimated 50,000 to 100,000 human genes might be patented in just a few years.

Predictably enough, this scenario provoked intense debate, involving a wide range of scientists, patent lawyers, biotechnology executives, policy makers, and others in the United States and beyond. Should patent offices recognize an EST as a patentable entity? If so, should a patent on an EST convey rights to the corresponding gene and protein? Critics of what became known as the "Venter patents" or the "NIH patents" thought that the claims overreached. (Despite Reid Adler's role in instigating the patent application, I do not recall hearing anyone refer to the "Adler patents.") The US Patent and Trademark Office (PTO) had previously granted gene patents, such as Genentech's patent on the gene for human insulin, in cases where the applicant had cloned a gene of known function. But the genes "indexed" by ESTs had not been cloned, and an EST by itself provided no information about biological function.[32] Nor did an EST indicate the gene's location in the genome. The NIH patent application thus raised what one commentator called "the patent question of the year": whether ESTs met the patent system's requirements of novelty, nonobviousness, and utility.[33]

The reaction from US scientists involved in the HGP was immediate and overwhelmingly negative.[34] Those scientists who regarded ESTs as a routine, not revolutionary, scientific development reacted with dismay. Speculation also began about the potential effects of EST patenting.[35] Scientists worried that patenting would delay the submission of cDNA sequences to the

Beyond the hassle of transporting clones, dependence on materials made it possible (in principle) for genome mappers to control access to maps for strategic reasons. Some gene hunters worried about the possibility that large genome centers would gain "unfair" advantages, perhaps by hoarding data or using their structural position to establish asymmetric collaborations. Gene-hunting laboratories, which anxiously sought data that would speed their search, also worried that centers might aid their competitors. "No one wanted," as the director of NORTHERN GENOME CENTER put it, to have to "suck up" to a genome center to obtain a map.[17]

Advocates of STSs saw the new landmarks as a solution to this problem. The reason: STSs—which relied on the recently commercialized polymerase chain reaction (PCR)—could be fully described in informational terms. Because STSs could be completely represented as written inscriptions, they eliminated the need to store and ship biomaterials.[18] An STS could be fully represented in terms of two short DNA sequences (each roughly two dozen base pairs in length) known as "PCR primers." As an example, consider the following text:

Forward primer: TCCTGGCTGAAGAGGTTCC

Reverse primer: CATTATAGGGGCCACTGGG

PCR product size: 192 base pairs,

which denotes an STS on human chromosome 22. Given this information, a molecular biology laboratory could use standard equipment and PCR techniques to test a DNA sample for the presence of this STS.[19] Laboratories could use STSs to make maps and to link maps made in different ways (figure 4.1). When hunting for a gene, an STS near the gene was extremely valuable.

Producing genome maps in a form that could be fully expressed as written inscriptions had important consequences for knowledge-control regimes because publishing a map and releasing the knowledge objects needed to use it became one and the same act. No one would have to obtain the underlying clones to use a map, and no one could publish a map and then fail to respond to requests for the underlying clones. Beyond making maps circulate more freely, STSs were also seen as a way to make genome centers more accountable. One of the architects of the STS proposal stressed that the advisory group's "single strongest motivation" for requiring STSs was to open up maps to systematic external evaluation. "We were looking databases and lead to increased secrecy in genome research.[36] Observers also predicted that the patents, if granted, would provoke "an international gene race that would make the Alaskan gold rush look sedate" and warned of an uproar "if a few developed countries, representing perhaps 10 percent to 15 percent of the human family, were to stake a preemptive claim to intellectual property rights over the human genome."[37] To such observers, the issues seemed too fundamental to be decided by the US PTO without broader international participation.

Within the NIH and the US government more broadly, EST patenting became a contentious issue. The NCHGR, itself a part of the NIH, strongly opposed the patent application. At the meeting in July 1991 where the filing of the application was first announced, NCHGR director James Watson, known for his impolitic remarks, objected that automated sequencing machines "could be run by monkeys" and argued that if EST patents could lock up entire genes, biomedical research would be stymied by endless patent litigation.[38] Watson's outburst captured two of the main complaints made by many HGP scientists: first, that awarding patents on whole genes for the straightforward task of sequencing cDNA fragments would grant too much intellectual property for too little creative scientific work and, second, that such patents would harm research and inhibit collaboration. In January 1992, the scientific advisory committee charged with guiding the NIH and DOE's genome programs issued a statement unanimously "deploring the decision to seek such patents," calling the claims "deleterious to science" and expressing concern that "the filing of such claims undermines the activities of the Human Genome Project." The committee also warned about the risk of an international "patent race" that could "compromise or destroy the international collaboration that we regard as essential for the work ahead."[39]

Struggles to control access to information about the patent applications developed within NIH. The scientific advisory committee called on the NIH to release documents for public comment, including the patent applications, which normally are kept confidential until patents are issued. NIH director Bernadine Healy, a supporter of the patent application, reportedly ordered Watson to refrain from further public comment on the matter.[40] Watson's clash with Healy over ESTs ultimately played a central role in his forced resignation in April 1992.[41]

Imagined Trajectories

The prospect of EST patents made epistemic claims about the expected trajectory of research using ESTs into performative speech acts with direct implications for the patent dispute.[42] How close was finding an EST to finding something of immediate value in pharmaceutical development, such as a gene with a known function related to a disease process? Supporters of the patents, such as Venter and Adler, depicted ESTs as having great epistemic value, imagining a research process that would progress straightforwardly from EST to full-length gene to protein. Going from EST to gene, in their view, might be so easy that finding the gene would become "obvious"—so without EST patents, genes might become unpatentable. In contrast, opponents of the patents depicted the process of moving from EST to full-length gene to determining biological function as highly contingent. "Premature" patents on genes might impose debilitating restrictions on future inventors, who would have little incentive to do the hard work of determining the function of genes that other parties owned.[43]

Most US genome scientists with whom I discussed the EST patent controversy supported the existing intellectual-property regime in biotechnology that encouraged the patenting of genes of known function. But they objected to granting a patent on a gene for the finding of a "mere" "tag." An interview with NATHAN, the director of the EASTERN GENOME CENTER, in February 1992 provides an example: "The usual arguments against patenting genes is that it's human patrimony, and we can't patent the human patrimony. Give me a break! … All sorts of compounds are the patrimony of nature. We might want to change, if we feel like it, but the patent laws seem to be fairly clear. Yes, you can patent the gene you fully describe." However, ESTs did not constitute a full description of the gene itself: "It's as if an organic chemist were to find a new drug in a fungus and were to characterize a prosthetic group hanging off it. 'There are two methyl groups and another group here. I claim the rest of the molecule.' … No composition of matter has been uniquely defined. … [It's] really [a] grant proposal for defining such a composition of matter."[44]

The "grant proposal" analogy frames an EST as a bit of preliminary data, not a significant finding, and defines identifying an EST as clearly a far cry from discovering a gene and determining its function. By no means were the critics saying that an EST had no value; scientists sometimes likened ESTs to markers used when hunting for genes. But although a marker known to flank a gene might be extremely valuable, finding such a marker was in no way considered equivalent to localizing and isolating the gene itself—a distinction painfully familiar to the disappointed losers of a number of races to clone genes for Mendelian disorders. Even so, the flanking marker analogy has some weaknesses. A flanking marker is "next to" a gene, whereas an EST is "part of" the gene and therefore might seem to provide a stronger claim to have "found" it.[45]

Configuration Power

Knowledge-control regimes, such as the patent system, play an important role in constituting agents with specific capacities and incapacities and thereby in allocating control among those agents. Granting a patent simultaneously produces several things: a patented entity, a patent holder, a residual population of nonowners, and a set of control relationships that allocate entitlements and burdens. A patent awards its owner the right to exclusive use of the patented entity and the privilege of making managerial decisions about how to exploit this right. Nonowners, for their part, are endowed with a duty to refrain from using the entity (absent permission), and they are accorded no right to interfere with the owner's managerial privileges.[46] Patent holders tend to draw from a familiar repertoire of strategic options, among them licensing rights to others (exclusively or nonexclusively), reserving use of the entity for themselves, using rights as a bargaining chip in negotiations with others, and blocking use of the entity entirely.

Beyond the market power emphasized in economic accounts of patenting, historical and social studies of science have shown how patents grant owners "configuration power"—the ability to influence how patented entities are fitted into sociotechnical forms of life.[47] Configuration power thus extends beyond such matters as price and access and encompasses the capacity to exert influence over the shape of emerging technologies and the social relations pertaining to them. On occasion, actors use configuration power to reinvent themselves, to reconfigure users, or to transform knowledge-control regimes. The free- and open-software movement, to take one important example, constituted a new kind of owner—one that uses copyrights, in conjunction with the innovative General Public License, to create control relationships that leave particular pieces of software usable by anyone while simultaneously blocking efforts to appropriate them and also allowing them to be improved.[48] This movement thus used configuration power to constitute a new knowledge-control regime with new kinds of agents, a new form of unownable property, and a novel system of control relationships.[49]

In the EST case, speculation abounded about what the patent holder (variously figured as "the NIH," "the US government," or "Venter") would do if the patents were issued. Would the government keep license fees nominal to stimulate research and development or try to generate revenue? Would the NIH reinvent itself as a central licensing authority, granting rights to use specific human genes in a manner loosely analogous to the way the Federal Communications Commission regulates the electromagnetic spectrum? Would it deploy licensing as a policy instrument to guide research agendas and product development? Would it license genes in ways that favored American companies over foreign ones?

NIH officials publicly argued that the decision to file EST patents was "not a final policy" but that they did not want to lose any rights while the issues were under discussion.[50] Likening patent practitioners to scientists who "formulate and test hypotheses," Adler called the application an "experiment" to find out whether ESTs were patentable and to determine if an EST patent would be tantamount to patenting the gene it indexed.[51] To observers not privy to the NIH decision process, the patent application seemed to be an ambiguous mixture of a test case to clarify the law, an interim measure while policy took shape, a preemptive strike to prevent patenting by others, and an attempt to capture intellectual property. Many found the NIH's public statements less than reassuring.[52] After all, the agency was aggressively claiming whole genes and proteins, filing applications on thousands of genes, refusing to withdraw the applications even under international criticism, and adjusting its patent claims to make them more likely to survive PTO scrutiny.[53]

International Contention

During the 1980s, Europe had been more cautious than the United States about expanding patents to cover life forms and biotechnological entities, and many European scientists and policy makers greeted the NIH patents with outrage.[54] Many specialists in US intellectual-property law doubted that the NIH's claims on full-length genes would hold up, but policy makers in other countries not surprisingly proved unwilling to count on the US

PTO to reject those claims. In the United Kingdom, the MRC, which had funded its Research Centre in Harrow to sequence cDNA clones for submission to public sequence databases, decided to stop public release of partial cDNAs. In June 1992, at the height of the EST patent controversy, I visited the United Kingdom and interviewed genome scientists and officials with the MRC. At the time, the MRC was preparing to file "defensive patents" on thousands of partial cDNA sequences that it had produced but had withheld from release.[55] My interlocutors quite uniformly saw the NIH application as damaging to genomics and believed that partial cDNA sequences should be in the public domain. One MRC official, active in shaping the United Kingdom's genome-mapping program, argued that EST patents were reintroducing "parochial" concerns with "ownership of your data," setting back international collaboration, and increasing "distrust."[56] UK scientists involved in cDNA sequencing expressed dismay. TREVOR, a young scientist involved in the UK project was upset that the United States was trying to "blanket commercialize" genes that it had not fully sequenced: "I have worked in a commercial company before and am only too aware that there is a necessity for commercialization of science. ... Now, if somebody has put an awful lot of work into a particular compound, ... then I think that is wholly appropriate that they should reap some benefit from the research and the finance that they have put into that. What I think is inappropriate is that somebody tries to do that with an entity that they know very little about." TREVOR remarked that the only positive thing to come out of the episode was something "frivolous": members of the British Parliament were actually talking about work that he himself had done.[57] Later that summer, the MRC filed patents on 1,200 cDNA tags.

*   *   *

In August 1992, the US PTO rejected the NIH patent applications for failing tests for novelty, nonobviousness, and utility, a move that provoked speculation about whether the decision was a fatal blow to EST patents. Healy sought to reverse the PTO's determination, and the NIH filed new patent applications on 4,448 ESTs.[58] The MRC continued its policy of defensive patenting. MRC officials said that they would be happy to make a "bonfire" with their applications, but doing so would be "premature" so long as the US threat remained.[59] The NIH's effort to patent ESTs was brought to a close by a policy change following the election of President Bill Clinton in November 1992. Clinton appointed a new NIH director, Harold Varmus, who withdrew all pending NIH patent applications in early 1994, thus diffusing the international controversy.[60] Legal wrangling about the patentability of ESTs continued for years, ultimately contributing to a policy change intended to toughen the utility requirements for patent eligibility.[61] EST patents were allowed, but the important claims—on full-length genes, proteins, and antibodies—were denied.

Object 4: Proprietary EST Databases

EST patents never became enforceable instruments for capturing control over large numbers of genes, but the concept of the EST patent and the prospect of quickly securing control over the genome helped inspire new vanguard visions of genomics as an immediate commercial opportunity. The US high-technology world—with its well-developed machinery for the capture of intellectual property, its institutionalized imaginary of the revolutionary startup company, and its ongoing demand for promising venture-capital investments—quickly made substantial investments in a wave of "genomics companies." Our fourth object—the proprietary EST database—became the centerpiece of two of these startup companies, both of which focused on selling genomic data. The expectation that genome research might begin producing treasure troves of intellectual property, not someday but soon, accelerated the "promissory economy" that Mike Fortun analyzes in his ethnographic investigation of the Icelandic company deCODE Genetics.[62] In the 1980s, Walter Gilbert's effort to found Genome Corporation had floundered for lack of financing. But in the early 1990s—especially in 1992 and 1993 after ESTs had drawn attention to genomics—venture capitalists partnered with genome researchers to launch startup companies. By early 1993, when *Science* reported that venture capitalists were "pouring money into genomics companies," about a dozen startups were in the works.[63] A report prepared for the US Office of Technology Assessment in 1994 discussed eight genome companies, estimating that they had collectively garnered some $160 million in capital, the "vast bulk" of which was raised in 1993, a year that was otherwise relatively bleak for biotechnology startups (table 5.1).[64]

All of the new genomics companies were founded to exploit the economic potential of high-throughput genomic technology. Like most biotechnology startups, they are best understood as research operations

Table 5.1

Selected Genomics Startup Companies Founded 1991–1993

Company	Year Founded	HGP Scientists Involved (selected)
Darwin Molecular	1993	David J. Galas; Leroy Hood
Human Genome Sciences	1992	Ronald W. Davis
Incyte Pharmaceuticals	1991	Charles R. Cantor
Mercator Genetics	1992	David E. Cox; Richard M. Myers; Eric D. Green
Millennium Pharmaceuticals	1993	Patrick Brown; Daniel Cohen; Eric S. Lander; Richard Wilson
Myriad Genetics	1991	Walter Gilbert; Jasper D. Rine; Mark H. Skolnik; Barbara Wold
Sequana Therapeutics	1993	Mark Boguski; Peter Goodfellow; David E. Housman; Hans Lehrach; Anthony Monaco

The companies listed can be grouped into two main types. "Positional cloning" companies (Mercator, Millennium, Myriad, and Sequana) sought to identify and clone genes involved in causing various common diseases, such as hypertension, asthma, and diabetes, that they had specifically targeted. Their main goal was to obtain patents on such genes for licensing to large pharmaceutical companies that sought to develop therapeutics. In contrast, the "database companies" (Human Genome Sciences and Incyte) built business models around proprietary EST databases. The final company, Darwin, planned to generate sequence information to use it in directed gene discovery. The founders, executives, and scientific advisory board members of these companies included leading scientists involved in the HGP. Most of these scientists remained active in the HGP and retained academic posts: Mark Boguski (National Center for Biotechnology Information), Patrick Brown (Stanford), Charles Cantor (Boston University, former principal scientist of the DOE genome project), Daniel Cohen (Généthon), David Cox (Stanford), Ronald Davis (Stanford), David Galas (former director of the DOE genome project), Walter Gilbert (Harvard), Peter Goodfellow (Oxford), Eric D. Green (Washington University), Leroy Hood (University of Washington), David Housman (MIT), Eric Lander (Whitehead Institute), Hans Lehrach (ICRF), Anthony Monaco (Oxford), Jasper D. Rine (Berkeley), Mark Skolnick (University of Utah), Richard Wilson (Washington University), and Barbara Wold (Caltech, chair of the NIH human genome grant-review panel). *Source:* Adapted from Cook-Deegan 1994b.

designed to produce intellectual property and profit from it either by commercializing it directly or, more often, by partnering with (or being purchased by) pharmaceutical or agribusiness companies with the resources to develop and market products. Two companies, Human Genome Sciences (HGS) and Incyte Pharmaceuticals, envisioned proprietary EST databases as the core asset of a biotechnology company.[65] Incyte sold nonexclusive access to its EST database to pharmaceutical companies, each of whom paid millions of dollars to subscribe in the hope of obtaining information about genes and proteins to target for drug-development research.[66] HGS, the better-capitalized startup, was soon being described as the "uncontested leader in the gene-hunting race."[67] Its complex business model involved founding not only HGS itself but also a nonprofit institute and establishing a set of alliances grounded in an elaborate assemblage of interorganizational control relationships. I call this assemblage the "HGS nexus" and concentrate on it in the next subsection.

The HGS Nexus

When the EST patent controversy transformed Venter from a little-known NIH researcher into one of the world's most visible genome scientists, he became a target of opportunity for venture capitalists.[68] In June 1992, Venter and HealthCare Ventures, an exceptionally well-capitalized venture firm, reached an agreement. The plan entailed founding two interconnected organizations—the for-profit company HGS and a nonprofit organization, The Institute for Genome Research (TIGR). Venter was made TIGR's president, chief executive officer, and chairman of the board.

A trio of press releases—from NIH, TIGR, and HGS—revealed elements of the plan in July 1992. NIH director Bernadine Healy announced that Venter was resigning his NIH position to join the private sector and presented his departure as an instance of successful technology transfer: "Now it is time for Dr. Venter to take his bold discoveries out of NIH, a great marketplace of ideas, and into the marketplace of American, private industry."[69] TIGR defined itself as a "new private, not-for-profit organization" with the mission of "accelerating the elucidation of the nature and function of human genes."[70] Initial funding for the institute would come from a $70 million, ten-year grant from Human Genome Sciences, Inc., a new company that would "collaborate" with TIGR as well as with pharmaceutical and biotechnology companies to "accelerate the development" of diagnostics and therapeutics.[71] TIGR explained that HGS "will have commercial rights to the institute's discoveries" but also that the two organizations intended to "freely offer" "all scientific data" to the NIH and the international research community.[72] The precise structure of the relevant control relationships remained unclear, given the ambiguity of such terms as *freely offer* and *collaborate*.

In engaging Venter, HGS benefited not only from his scientific know-how and reputational resources but also from the fact that Venter brought the bulk of his scientific team from NIH to TIGR.[73] "Technology transfer" in this instance enabled the HGS nexus to acquire a key holding of Venter's NIH laboratory: an already assembled research group with the specific constellation of skills and knowledge needed for high-throughput EST production. Coupled with ample funding, these assets enabled HGS and TIGR to scale up rapidly, building a sociotechnical system of sequencing machines, bioinformatics tools, and personnel that soon had generated tens of thousands of ESTs.[74]

For about year after the formation of TIGR and HGS, how the company planned to make a profit remained a matter of speculation among HGP scientists. As an example of how people made semi-educated guesses, consider a conversation at the Cold Spring Harbor meeting in May 1993. I was eating lunch with some scientists from an NIH-funded genome center when we were joined by SANGSU, a scientist who presented himself as having relatively solid information about TIGR's activities.

Sangsu said that Venter's sequencing operation was getting 500 unique cDNAs per day.

"Hmmm," the center director said. The group began running through the numbers aloud: "That's plausible, that's not bad; let's see, they have 32—is it 32 or 36?—ABI machines. So let's see, yeah, that's plausible because if they ran them full, they could get 1,000 per day. So that's not even that high if they are getting 500 unique ones."

At one point, the center director said, "I don't understand how that company, HGS, is going to make money. I don't understand what their plan is." Another scientist asked, "Well, are they still counting on cDNA patents?" The center director said, "I just don't think that they could be that sure about those patents, so there has to be more than just those patents in their plan."

"Yes," Sangsu said, "especially if they stick to that nine-month policy for the release of data."

"They promised to release the data in nine months?" one scientist asked. "Yes, they have."

"Very interesting," said the center director. But that new (how new?) bit of (reliable?) information seemed to leave him, if anything, even more uncertain.

Later that month the broad outlines of the HGS business model became visible when HGS announced an alliance with the pharmaceutical giant SmithKline Beecham. SmithKline was to provide $125 million to HGS to conduct large-scale genome sequencing. The terms were not fully disclosed, but news accounts indicated that the companies would use sequence data "to develop new drugs and diagnostic products"; that SmithKline would have exclusive rights to most products and services developed; and that HGS would receive royalties.[75] The move was widely interpreted as an attempt to leap ahead of competing pharmaceutical companies in genomics.[76]

More information emerged in connection with HGS's initial public offering (IPO) of shares on the NASDAQ stock exchange in December 1993. TIGR and HGS were creating a huge proprietary EST database, with which they hoped to "find" or "identify" the majority of human genes. HGS also planned to identify and characterize genes of bacteria, fungi, viruses, and plants. The company would "fully sequence those genes that appear most likely to have commercial value."[77] It would protect its EST database—its most important form of what Kaushik Sunder Rajan and others have termed "biocapital"[78]—using trade secrecy. In contrast to Incyte's strategy of selling multiple subscriptions, HGS would use its database in its own research and to leverage collaborations. Its core strategy was to patent a small number of fully sequenced genes chosen for their role in human disease, although it did file EST patents claiming entire genes (38,000 of them by the end of 1993).

HGS's Security and Exchange Commission filings outlined the control relationships underlying the HGS–TIGR–SmithKline nexus.[79] HGS and TIGR had entered into a Research Services Agreement and an Intellectual Property Agreement under which HGS was to provide TIGR with a total of $85 million over ten years. HGS also gave 1,609,884 shares of its stock to TIGR, which distributed them to Venter and other TIGR employees. In exchange, TIGR was to disclose to HGS "all significant developments relating to information or inventions discovered at TIGR." HGS also would own royalty-free rights to inventions and patents arising from TIGR's research.[80] TIGR, a nonprofit organization, thus would be producing intellectual property for a startup company.

Under the legalistic definitions of corporate law, HGS and TIGR were "independent" entities, but TIGR also gave HGS some rights over its research activities.[81] TIGR agreed that it would sequence human genes selected by a committee composed of scientists from TIGR and the HGS; operate eight DNA-sequencing machines "at the Company's direction"; and delay publication of DNA sequences to give the company an opportunity to exploit them.[82] The delays—up to eighteen months after initial disclosure to HGS—assured the company a period of exclusive access to TIGR data.[83]

HGS's IPO-related documents also provide a look at the Collaboration Agreement between HGS and SmithKline Beecham. The agreement required HGS to disclose to SmithKline all of its "Human Gene Technology"—a category encompassing "data, material, know-how, and inventions." It also granted SmithKline the option, grounded in the legal mechanism of a right of first refusal, to pursue research and development (R&D) based on HGS human gene technology in a domain defined as "the SB Field." The SB Field included the entirety of human and animal health care, with certain specified exceptions, such as gene therapy and antisense products. HGS granted SmithKline exclusive worldwide rights to "make, use and sell" any products resulting from this R&D. In exchange, SmithKline paid HGS $22 million, purchased 1,012,673 shares of HGS stock for $37 million, and agreed to pay more as certain R&D "milestones" were reached. Milestone III, to give one example, would be reached when HGS transferred 90,000 ESTs to SmithKline, prompting a payment of $25 million before May 1996 or $20 million after that date. SmithKline also agreed to pay royalties to HGS on any products resulting from their alliance.[84]

The Collaboration Agreement also constituted a "research committee" with equal membership from SmithKline and HGS.[85] This committee was entitled to direct the research performed using 35 sequencing machines and 15 robots operated by HGS. It was also charged with approving "research programs" in the SB Field. The Collaboration Agreement also allocated the privilege to pursue "approved research programs" between the two companies. For example, SmithKline would be given 90 days to decide whether to undertake or reject a research program, and HGS retained the right to pursue any program that SmithKline chose to reject. The agreement also stipulated that each company would pay royalties to the other on any products resulting from their collaboration.[86]

The contractual machinery described here thus constituted an assemblage of organizations that, although legally independent corporate entities, were tightly linked through carefully choreographed exchanges of money, data, and technology. The resulting knowledge-control regime also allocated control over the research process itself, aiming to unambiguously specify the entitlements of each organization while leaving some flexibility for moving in different directions if their goals diverged.

Query-Based Collaborations

HGS also constituted collaborations with university-based researchers, especially after early 1994, when most biologists, "even former skeptics," concluded that querying EST databases was quite useful.[87] HGS used its collection of ESTs—or, more precisely, its capacity to query its databases and selectively distribute results—to leverage alliances. Using the mechanism of the Material Transfer Agreement, HGS began establishing contracts allowing academic researchers to search for specific sequences in its EST database in exchange for granting the company exclusive rights in any discoveries.

A scientific success in 1994 convinced many scientists that EST databases were valuable. Bert Vogelstein, a cancer researcher at Johns Hopkins University, entered into a query-based collaboration with TIGR and HGS to look for a gene implicated in colon cancer.[88] The query identified a gene, and in February 1994 HGS announced that Johns Hopkins had signed a Material Transfer Agreement granting the company an exclusive license.[89] A paper published in *Science* in March 1994 described the gene, with scientists from Vogelstein's group, HGS, and TIGR all named as authors.[90] News of the upcoming publication "apparently leaked out" a day early, causing HGS's stock price to jump from $15 to $17.25.[91] HGS filed a patent application on the gene. News coverage in the scientific and financial press persuaded many observers that databases containing lots of ESTs were a powerful tool for gene discovery.[92] The speed with which the Vogelstein group had found the gene proved especially impressive to scientists worried about being scooped by competitors.[93] HGS held up the collaboration as a model for how HGS plans to work with investigators in not-for-profit institutions.

Query-based collaborations thus came to occupy a new position in the strategic imagination of gene hunters in academic institutions. By June 1994, HGS had signed 30 research agreements with 19 different institutions, including Boston University, the Mayo Clinic, McGill University, Emory University, and the ICRF in London.[94] Exclusive control of its EST database made HGS into an "obligatory passage point," presenting disease-gene hunters with a straightforward choice: forgo access to this promising database or agree to terms granting HGS (and ultimately SmithKline owing to the layering of agreements) the option to license any patents.[95] HGS and TIGR were also entitled to preview forthcoming publications. The HGS database thus provided the company with a tool for drawing university researchers, along with some of their valuable knowledge objects, into collaborations that might yield exclusive rights. In effect, the company was incorporating an additional organizational layer—consisting of university researchers embedded in carefully codified control relationships—into the HGS nexus. Through such a knowledge-control regime, a company can extend its perimeter in ways that capture elements of a university's research capacity.

HGS's strategy of using its EST database to leverage query-based collaborations in exchange for patent rights met some resistance from academic scientists.[96] Even some scientists deeply involved in commercial genomics regarded HGS's demand of patent rights as intrusive and, as one such researcher put it, a bit "nasty."[97] More broadly, some scientists contended that HGS would slow the advance of science because restricting access to the database inevitably would lower the number of scientists—in both academia and industry—who would make use of this valuable "research tool." To many HGP scientists, it seemed self-evident that ESTs should be housed in databases dedicated to providing unencumbered access, such as GenBank and the European Molecular Biology Laboratory (EMBL) database. These scientists were committed to constituting sequence data as an open-access resource—a vision inscribed into the governing frame of the US genome program, as we saw in chapter 4, and strongly supported in the UK and on the European continent. They thus opposed business models, like those of HGS and Incyte, that depended on "locking up" or "hoarding" the "fundamental" data about the human genome—namely, the genome sequence "itself," a category into which they assimilated partial cDNA sequences.

Such objections were not directed at all forms of commercial genomics but at the specific approach taken by HGS and Incyte. A number of HGP scientists were involved in positional cloning companies, such as Myriad Genetics, Mercator Genetics, and Sequana Therapeutics (table 5.1), that sought to identify and clone specific disease genes. The possibilities of such companies, along with the politics of the knowledge-control regimes that they might constitute, came into sharp view in 1994 when Myriad Genetics and the University of Utah found the first gene implicated in a common human disease: breast cancer. The parties filed for a patent on the gene, BRCA1, which plays a role in some cases of breast cancer.[98] (The PTO did grant the patent, and extensive controversy thereafter surrounded Myriad's use of BRCA patents to obtain monopolistic control over genetic testing for breast cancer and important data for breast cancer research.[99]) Whereas some HGP scientists were critical of disease-gene patents, many others supported them but at the same time opposed the development of proprietary EST databases. Finding disease genes was much more difficult than generating ESTs, they reasoned, so perhaps those who isolated and cloned a gene deserved a patent, and patents might be necessary to stimulate companies to develop therapeutics. In contrast, ESTs were "research tools," not discoveries of something with actual medical value, so they should be made available to all scientists. KEVIN, a scientist who was critical of the HGS nexus and an adviser to a positional cloning startup, illustrates this way of thinking. "Maybe I'm biased because I'm involved in it," Kevin told me in 1994, but "I just have a lot less problems with" the gene-cloning companies than with the database companies HGS and Incyte. The "worst thing you can say" about the gene-cloning companies, "is, yes, they probably take some talent out from the academic genome areas, but at least it's fundamentally not this blatant thing of just creating these huge databases of information and then not releasing them."[100]

Hostility to business models dependent on restricting access to EST databases was palpable at the Cold Spring Harbor meeting in 1994, held less than two months after the news broke about identifying the Vogelstein colon cancer gene. To get a sense of the mood at the meeting, consider the following episode, which took place during a plenary session with hundreds of people in the room.

MATT, a scientist from TIGR, gave a talk about its EST database, which he said contained 115,000 ESTs. He also presented what he described as some very preliminary estimates of the number of human genes. He did the estimates using four methods, each of which predicted between 60,000 and 70,000. During the Q&A, a GENOME CENTER DIRECTOR attacked Matt in a style reminiscent of a trial lawyer.

GENOME CENTER DIRECTOR: This is an impressive collection of data, very valuable data, something that represents several years' of work. How long did it take to produce it, 18 to 20 months, right?

MATT: Well, the facility really reached operation about a year ago. There was a ramping-up period.

GENOME CENTER DIRECTOR: Yes, so this is a substantial amount of work, and this is a resource of terrific value. So the question I have for you is, When will the data be made public?

The room got very still.

MATT: In the fall of 1994.

GENOME CENTER DIRECTOR: You know, this collection represents a lot of work. You've been talking about it for a long time. And I remember that about a year ago we were promised that it would be released a year later. So when is it coming?

MATT: Well, right now the informatics group actually has to do the task of getting this information packaged and out before September 1994, so we're moving against that wall at 60 miles per hour.

GENOME CENTER DIRECTOR: Surely, it's trivial for an informatics group that can organize all of these databases to simply post this information, so what is the problem?

There was some applause.

MATT: Well, the main constraint right now is all of this work was done with nonpublic money.

The next questioner, VICTOR, remarked: It's not quite true that all this work was done with nonpublic money because you made extensive use of public databases, such as Genbank, and incorporated information from them into your analysis.

Then Victor paused for a moment and went on to ask an unrelated question.

Victor's challenge to the idea that "all of this work" was done with private funds highlights an important issue: most of the uses of an EST database depend on having other sequence databases—the more comprehensive, the better—to use in conjunction with it, and the world's most complete sequence databases were the open-access ones, such as GenBank. Just as connecting a private drive to an extensive network of public roads enhances its value, using a private EST database in conjunction with these extensive public databases makes it more useful.[101] EST research thus depended on assemblages of "private" and "public" resources. TIGR and HGS were controlling access to the proprietary part of this assemblage (their unique EST databases) while making use of the open-access part (GenBank). The broader point, however, is that in contemporary biomedical science, attributing a research result to either "public" or "private" financing is an oversimplification of a process that typically involves resources, actors, and action embedded in a variety of jurisdictional spaces and interconnected knowledge-control regimes. The HGS nexus—with its assemblage of agreements linking the company to a nonprofit institute, a multinational pharmaceutical company, and multiple universities in several countries—is a case in point.

Object 5: A Privately Funded, Public EST Database

Our fifth object—a privately funded, public EST database—was created in response to the HGS business model. The NCHGR leadership opposed using the HGS database in genome project work, and HGP scientists partnered with the pharmaceutical company Merck, a competitor of Smith-Kline, to independently generate vast numbers of ESTs and establish an open-access EST database.[102] In September 1994, Merck announced its intention to fund the production of a "public resource" of EST data to be known as the Merck Gene Index. Observers read the move as a way for Merck to provide university researchers with an EST database that they could query without getting entangled in agreements with SmithKline, an important competitor.

The Merck Gene Index was the first instance of a family of regimes that grew important in genomics—namely, large projects to produce genome sequence data for public release that are funded by private, profit-making enterprises.[103] With Merck funding, Robert Waterston and Richard Wilson, who were running a successful sequencing project at Washington University, launched a large-scale program to generate ESTs. Washington

University was able to begin EST production quickly because, as Venter had recognized, the shift from genomic sequencing to EST production was mostly a matter of sequencing different DNA. A Columbia University laboratory provided cDNA libraries, and the Lawrence Livermore National Laboratory arrayed them in microtiter plates for sequencing.[104] Washington University sent the sequences to GenBank for inclusion in dbEST, a database devoted to ESTs.[105] In February 1995, the project submitted its first 15,000 ESTs—a number expected to reach 300,000 within 18 months. Researchers had soon queried GenBank's EST database some 150,000 times.[106]

Like the rest of GenBank, dbEST made its information available to any and all with no strings attached. Merck researchers did not get early access to the information. Merck's statements to the news media stressed scientific progress, arguing that unrestricted access to ESTs would increase the probability of discovery.[107] HGP scientists were grateful. An allegorical drawing on the abstract book for the Cold Spring Harbor meeting in 1995 figured Merck as a white knight saving genes from being imprisoned in a dark castle marked with an H.

Object 6: ESTs as Ordinary Tools

The Merck Gene Index made it possible for university researchers to access an EST database without acceding to the control relationships that HGS was imposing on its query-based collaborations. dbEST grew rapidly, and a large and growing number of scientists were soon integrating ESTs into their research programs.[108] With dbEST offering similar data for free and without restrictions, HGS's database grew increasingly irrelevant. Moreover, as the combined total of human ESTs accumulating in the databases came to greatly exceed the total number of human genes, researchers concluded that the early worries about redundancy in cDNA libraries had proven well founded: each gene was being tagged by many ESTs, so the EST databases were cluttered with tags for the same genes. Because ESTs were based on partial cDNAs, two ESTs that indexed the same gene would often not overlap, making it look as if each EST indexed a different gene. Redundancy thus significantly reduced the value of an "index" of genes based on ESTs.

These developments transformed the EST into the sixth and final object of interest here—an "ordinary tool." EST databases that were both incomplete and highly redundant did not amount to a shortcut to "finding" all the human genes, as Venter had imagined. Patent applications that did not survive scrutiny could not yield a treasure trove of intellectual property. Proprietary EST databases could not squeeze scientists into collaborations with burdensome terms if the data could be accessed elsewhere without such restrictions. To be sure, ESTs remained useful, but they were no longer objects with the potential to radically reorder the world of the genome.

In the aftermath of this deflation of the value of ESTs, HGS and Incyte began shifting their business models to other areas. Venter shifted TIGR's sequencing capacity to a new target: the genome of a bacterium, *Haemophilus influenzae*, and TIGR became the first organization to completely sequence a free-standing organism—the ambitious goal that Walter Gilbert's laboratory had attempted a few years earlier in an unsuccessful pilot sequencing project.[109] Venter, whose relationship with HGS had grown strained, severed TIGR's relationship with the company in 1997, giving up substantial funding to decouple TIGR from the HGS nexus.[110]

The new status of ESTs as an ordinary tool was nicely captured in an interview with Lew, director of COASTAL GENOME CENTER, in January 1996. Speaking about a year after the first Merck-funded EST sequences started being transferred to dbEST, he argued that the "cDNA-sequencing projects" that had made their data public "have had a big effect on biology," making the hunt for disease genes "much, much faster now." Lew was excited because his laboratory had just identified a disease gene, but despite his enthusiasm about EST databases he clearly saw "little snippets of cDNA" as a useful yet imperfect tool. For him, cDNA sequencing had always been a short-term strategy, best thought of as part of the mapping phase of the genome project—which he expected to end soon. We spent much of the interview discussing his views of the challenge of shifting the focus of the HGP from mapping to sequencing—a problem that captured his scientific and strategic imagination. For Lew, ESTs had become a mundane part of the genomics tool kit.[111]

Conclusion

This chapter has examined the contested place of partial cDNA sequences in the world of genomics between the late 1980s and roughly 1996. Through a contentious process of ongoing change, a series of distinct knowledge objects took shape: namely, the cDNA strategy, the EST, the EST patent, the proprietary EST database, the public EST database, and finally the EST as an ordinary tool. During this process, the epistemic and strategic value of partial cDNA sequences proved radically unstable. This instability was not merely a matter of cognitive reinterpretation but involved transformations of strategic objectives, material practices, organizational structures, and, not least, knowledge-control regimes. Objects, objectives, and regimes were simultaneously reconfigured as actors pursued new strategies for constituting knowledge and colonizing futures.

As they were reimagined and repackaged into new objects, partial cDNA sequences were radically transformed. To be sure, the array of base pairs in any particular cDNA sequence is not altered when it is imagined as an efficient shortcut to finding all the genes, included in a patent application, entered into a proprietary database, or reduced to the status of an ordinary tool. But in an important sense—indeed, in *most* important senses—these sequences became entirely different things during the action. The struggles surrounding these objects entailed significant change. Genomics companies were founded and funded. New actors such as HGS and Incyte joined the genomics vanguard. Powerful institutions such as Merck took on new roles. Along the way, cDNA-related objects and control relationships experienced repeated transformations, sometimes dramatic and sometimes subtle, as a process of interactive coproduction reordered strategies, practices, and knowledge-control regimes.

A final question remains, however: can any pattern be discerned in the turbulent controversy described above? Indeed, comparing the changes actors envisioned with the outcomes observed reveals a rather strong pattern: the most disruptive of the upheavals that actors envisioned did not take place. EST patents did not suddenly allocate control over the majority of human genes, and the NIH did not transform itself into a licensor of access to the genome. HGS did not succeed in entangling large numbers of academic researchers into collaborative alliances, and SmithKline did not leverage a short-term lead in EST production into a large and lasting advantage over competing pharmaceutical companies. The HGP goal of making sequence data openly available remained both policy and practice. And the PTO denied claims to entire genes based on ESTs, thus greatly reducing the value of EST patents while preserving existing biotechnology patent policy, which allowed patenting of cloned genes of known function. In short, none of the most radical possibilities materialized, even as the world of genomics was significantly transformed.

6 Regime Change and Interaction

Political revolutions sometimes produce lasting periods of ongoing instability as new regimes attempt to establish internal control and constitute orderly relations with external powers. Regimes may rise, fall, and reemerge in rapid succession. Boundaries may be redrawn, long-standing settlements may be contested anew, and disorder may even spread to adjacent lands. This chapter examines how periods of scientific change can produce an analogous, if less dramatic, dynamic of destabilization and reconfiguration, sometimes resulting not only in the formation of new knowledge-control regimes but also in adjustments to well-established ones. During such periods, how are control relationships reordered as regimes interact at points of contact? To what extent and in what ways are jurisdictions redrawn? And how do some changes come to span borders whereas others are largely or completely contained on one side?

To explore the process of regime change and interaction, this chapter focuses on knowledge-control regimes associated with DNA sequence databases—especially the US database, GenBank—and large-scale sequencing in the context of the HGP. Beginning in the early 1980s, GenBank (among other biomolecular databases) grew increasingly important to many fields of biology, serving as a complex hybrid of centralized information repository, scientific instrument, communication system, and arena for constructing and coordinating scientific ontologies. Building a sociotechnical system for collecting and making available an exponentially increasing amount of sequence data raised many challenges, not least the problem of governing what amounted to a new form of scientific communication. How should new regimes built for databases articulate with "adjacent" regimes, especially the ones governing the journal and the laboratory? How should jurisdictional questions arising at points of regime contact be addressed? And to what extent should the traditional control relationships of these adjacent regimes be adjusted?

To capture change over time, my account begins in the late 1970s, when DNA sequence databases were first established, and ends in 2003, when the HGP was officially completed. During this period, knowledge-control regimes were highly unstable, and I examine a series of five regimes associated with GenBank that took shape in succession, discussing the emergence of each, the problems and controversies that destabilized the first four, and the changes in control relationships intended to address those problems. The chapter analyzes how these new regimes interacted with established ones, such as the regime of scientific journal publication and the regime of the laboratory, and offers an account of the dynamics and scope of change.

Regime 1: Staff-Driven Collecting (1979 to the early 1990s)

Several historians, notably Hallam Stevens, Bruno Strasser, and Timothy Lenoir, have examined the rise of sequence databases and their place in making new kinds of biological knowledge.[1] In the late 1970s, Frederick Sanger in the United Kingdom and Allan Maxam and Walter Gilbert in the United States developed practicable methods for sequencing DNA, and an increasing number of papers containing DNA sequences began appearing in the literature. Gilbert and Sanger won the Nobel Prize for their contributions to sequencing methods. As the 1980s progressed, biologists increasingly regarded sequencing as a "technical" procedure rather than a "scientific" accomplishment. Most scientists who sequenced DNA typically did so in connection with specific biological research projects that required sequencing as a step in a broader analysis of biological functions or mechanisms. DNA sequences typically served as data—often the crucial data—for supporting knowledge claims.[2]

In the 1970s, the volume of sequence data in the published literature was tiny, but as data began to accumulate, a few research groups grew interested in gathering together the available sequences—from all organisms— and analyzing it using computational methods.[3] At the Los Alamos National Laboratory, a group led by physicist Walter Goad began to collect DNA sequences from scientific journals and to develop software for managing, searching, and analyzing these data. This project became the Los Alamos

Sequence Library (LASL), one of the first concerted efforts to create a centralized collection of biomolecular data.[4] Goad envisioned a carefully cataloged collection that scientists could consult and that would make sequences freely available for computational analysis. In 1982, Goad's group won federal support, and LASL became GenBank, a nationally chartered, federally funded database, with the goals of providing an archive of sequence data equipped to serve as a research platform for examining and analyzing sequences.[5] In Europe and Japan, science-funding agencies also established DNA sequence databases, the EMBL Nucleotide Sequence Data Library and the DNA Data Bank of Japan.

LASL built its database using the method of staff-driven collecting: employees searched the scientific literature for papers containing nucleic acid sequences, identified the sequences and other relevant information, and entered them manually. After LASL became GenBank, it continued to employ this approach for several years as it worked to collect "all published sequences over 50 nucleotides in length."[6] Each GenBank entry positioned the sequence in biological and bibliographic space, providing a journal citation and an "annotation" describing the source of the sequence (e.g., the organism or gene) and its "internal features" (e.g., the locations of protein-coding regions).[7] The database—essentially a collection extracted from the literature—was established quickly, expanded rapidly, and soon became indispensable to many biologists, growing in utility as it grew in size.

GenBank established a knowledge-control regime that constituted sequence data as open to any and all, making them available on magnetic tape and online with no restrictions on use or redistribution.[8] Like bibliographic index services, this incarnation of GenBank was an "add on" that LASL constructed adjacent to the knowledge-control regime governing journals. Collecting was conducted independently of journals, requiring no changes in how they operated. Journals continued to evaluate manuscripts and accept those that they deemed worthy, and they did not object to GenBank redistributing sequences from published papers. Authors were also minimally affected. Not only did the journals establish their priority, but most authors also considered the published paper, not the sequence data itself, to be the important source of credit. In short, the regime of staff-driven collecting did not impinge on any of the traditional control relationships of the journal publication regime. Nor did setting it up involve significant changes in relations with the laboratories that published sequences.

The First Destabilization: Exponential Growth

Staff-driven collecting was highly labor intensive. GenBank staff had to find sequences in the literature and enter them into computers. To ensure accuracy, they typed each sequence twice. The most time-consuming task was preparing annotations because describing the biological significance of a sequence required reading the paper in which it appeared.[9] GenBank soon found it could not keep up with the accelerating pace of sequencing. As a stopgap measure, GenBank, EMBL, and the Data Bank of Japan divided up the journals, each taking a share and exchanging data. The explosion of sequence data soon overwhelmed even this move, and GenBank was forced to sacrifice some combination of timeliness, accuracy, quality of annotation, and completeness of coverage of the literature.[10] Looking back on this period from 1991, BRIAN, one of the architects of GenBank, explained: "It is wrong to think that one central group can know all about the data and arrange it properly and manage it well. In GenBank, we tried that. ... You just can't do it. You can't have enough people with enough expertise to do that centrally."[11]

In 1985, GenBank began entering new sequences without internal annotations, which it planned to add later. This solution reduced the backlog of unentered sequences, but some users objected to the lack of annotation.[12] One database manager, DAVE, recalled that as delays lengthened, some biologists grew obsessed with timely data entry.

So they start to get it in their head that timely is important, ... [and] you say, "Well, there is a trade-off. The faster I bring it in the database, the less time I've got to check it for validity and that sort of thing. ... What level of error are you willing to tolerate?" ... And from some biologists I was getting the answer, "There is no trade-off point. Timeliness is the only factor." I said, "That is crap. That is pure crap. If timeliness is the only answer, I'll make up sequences. You can have them before the research is done. ... Come on! What's the real trade-off?" I couldn't get an answer out of them.[13]

These frustrations reflect the untenable position of the staff-driven collecting regime in a fast-moving, competitive field in which scientists wanted access to their colleagues' data immediately after publication, not months later. By the second half of the 1980s, a sense of "crisis" dominated the workshops and advisory committee meetings where the future of sequence databases was discussed.[14]

As the number of papers relying on sequence data skyrocketed, another problem arose: journals became increasingly reluctant to publish sequences owing to spatial constraints. Sequences, originally considered an important *part* of the scientific papers that drew on them, began to be seen as no longer belonging in the paper itself. No scientific paper can present a full account of the data or procedures that underlie it, and conventions about precisely what merits inclusion change over time.[15] As journals began to limit the space devoted to printing sequences, authors responded by omitting "uninteresting" sequences, such as data on noncoding regions (or what supporters of the cDNA strategy referred to as "junk"). However, this trend threatened some lines of research that needed the omitted data.[16]

Nucleic Acids Research (*NAR*), a particularly heavy publisher of sequences, provides a good example of how criteria for publishing sequence data evolved. Richard T. Walker, an editor of the journal since its launch in 1974, described in 1989 how *NAR* had initially accepted almost any DNA sequence and insisted that the "primary data" (namely, autoradiograms of the sequencing gels) be published or at least made available for peer review. Managing the flow of autoradiograms soon became a nuisance, however, so referees stopped checking the interpretation of the gel images. *NAR* then began refusing papers that "did not give the full sequence of a gene or cDNA." Next followed "a ban on publication of a gene sequence from one organism where the same gene sequence was already available from a related organism." By 1987, Walker explained, *NAR* would "only accept full sequence papers in which the product of the gene is relevant to nucleic acid biochemistry." The journal discouraged all "sequence-only" papers, favoring those using sequence data as "the starting point for the main thrust of a paper that provides new biochemical information."[17]

Journals also took steps to pack more sequence data into the same amount of space. *NAR* developed a "For the Record" section so scientists could publish "complete but 'uninteresting' sequence in the minimum possible space."[18] A tiny font allowed 14,121 nucleotides to fit into just three pages.[19] The *Journal of Biological Chemistry* tried a different approach, presenting DNA sequences in a compact format "directly readable from the journal page by a desktop computer." The sequence could be immediately transferred—"in a utilizable form, with annotation"—to the scientist's computer.[20] Such measures could not really address the fundamental problem of exponential growth in sequence production, though. Even in the *NAR* "For the Record" format, for instance, the 3 billion nucleotides of the human genome would occupy more than 600,000 pages. Database managers and journal editors alike were convinced that an entirely new approach was needed.[21]

Regime 2: Mandatory Direct Submission (1988 to the present)

A transition to a new knowledge-control regime took place in the second half of the 1980s as the EMBL Data Library and GenBank worked to develop a system through which scientists would submit sequences directly to the database and annotate their own data. In contrast to staff-driven collection from published papers, which the databases were able to implement by themselves, a regime that made direct submission mandatory required restructuring the burdens and entitlements of journals and authors.

The attractions of getting authors to submit data directly had begun to be discussed even before GenBank was officially chartered.[22] Indeed, almost from its inception GenBank encouraged scientists to send machine-readable versions of their sequences directly to Los Alamos, and by 1987 "a substantial amount" of data arrived either through electronic mail or on magnetic tapes and floppy disks.[23] Direct submission remained voluntary and, from the point of view of the databases, massively underutilized. EMBL reported that only about 35 percent of authors complied. Moreover, fully 70 percent of those who did comply sent their data as computer printouts rather than in machine-readable form, which meant data bank staff had to retype sequences that clearly already resided in the researcher's computer.[24]

Part of the problem was that the academic reward structures that motivated submissions to journals did not apply to database submissions, and no one expected tenure and promotion committees to treat contributions to sequence databases as significant scientific accomplishments. If anything, there were disincentives for submitting data. Not only was preparing and formatting data for transfer to the database a time sink, but an intentional strategy of delayed release could also prolong a comparative advantage. *NAR* editor Walker argued that researchers "have little incentive at present to deposit the data quickly—indeed, it can sometimes be an advantage to delay deposition. Scientists would like access to everyone else's data though they do not necessarily wish to reciprocate."[25]

Because sequence databases could not simply wish direct submission into existence, they needed to induce scientists to submit their data, and they needed to equip them with tools that would enable them to prepare acceptable submissions and usable annotations. A heterogeneous set of means was used to accomplish this change of regimes.[26] One crucial task was building an easy-to-use system for processing direct submissions, featuring "friendly if not seductive annotation software."[27] GenBank introduced an online interface, AUTHORIN, which was designed to ease submission and automatically check for common errors, permitting authors to correct draft submissions without involving database staff.[28] GenBank later added the Annotator's Work Bench—software to help submitters annotate their sequences.[29] After the release of Mosaic, an early and influential web browser, in 1993, GenBank developed BankIt, a new web-based "data submission tool."[30]

Even more significantly, the databases, along with their allies in the funding agencies, began trying to persuade journal editors to *require* authors to send sequence data to the databases prior to the publication of a paper. Competition among journals for the best scientific papers complicated these negotiations. Journal editors recognized the importance of making the sequence data available and saw direct submission as a way to address the problem, but with both reputation and revenue at stake they worried that if a journal required authors to submit sequence data prior to publication, authors would send their papers elsewhere.[31] Moreover, given their record of backlogs, the databases had trouble persuading editors that delays would disappear. Funding-agency officials served as intermediaries, assuring journal editors that the databases would not "screw up" and cause troublesome delays.[32]

The first journal to make publication contingent on database submission was *NAR*, which began requiring authors to send their sequences to EMBL on January 1, 1988.[33] The announcement of the new policy explained that *NAR* authors would be "guinea-pigs in an experiment" to try to cope with the "expected flood" of sequence data. Authors would have to contact EMBL and get an accession number "BEFORE submitting the manuscript." However, the announcement continued, "EMBL has assured us that accession numbers can be allocated very quickly and potential *Nucleic Acids Research* authors will receive priority. Anyone who experiences any delay should contact oné of the Executive Editors immediately."[34] Beyond documenting that submission had occurred, the requirement that authors had to provide accession numbers had a second advantage: including an accession number in the published paper provided a previously unavailable pointer *from* the printed articles *to* the database entries, thus tying the information in the journals and the databases together more firmly through bidirectional links.

By 1992, at least 36 natural science journals required sequences to be submitted to the databases as a condition of publication.[35] A full transition to direct submission could not be implemented instantly, however, not only because journals did not adopt direct submission at the same time but also because staff-driven collection had missed sequences during the 1980s that GenBank wanted to capture.[36] Moreover, not all journals welcomed mandatory submission. The most prominent opponent was *Nature* editor John Maddox, who debated supporters of mandatory submission in the journal's editorial pages in an exchange in 1989. Maddox saw the sequence databanks as important to research, and *Nature* had a policy of urging voluntary submission. However, he saw imposing a submission requirement on authors as an inappropriate addition to the proper role of the journal, which was to communicate scientific findings and maintain quality through peer review. Journals, he wrote, have "the right to ask for supporting data (sequences, atomic coordinates) even when they have no space to publish them." But they "have no right to adjudicate upon a contributor's subsequent conduct—and have few sanctions anyway. If there must be policemen, grant-making agencies are better placed." For Maddox, the journal's jurisdiction extended only as far as the author's manuscript and the data directly supporting it. Under his logic, authors were obligated only to provide to the journal precisely those parts of a sequence needed to buttress a knowledge claim, and they retained the privilege to keep the rest secret— something he saw as especially relevant to commercial organizations, which "are notoriously unwilling to let their competitors know what interests them, and may legitimately claim the right to secrecy while enjoying free access to what academics publish."[37]

Supporters of mandating submission strenuously objected to these arguments. Three representatives of EMBL argued that the integrity of the publication process requires submission, for "readers cannot assess conclusions based on data they cannot see."[38] An executive editor of *NAR*, Richard J. Roberts of the Cold Spring Harbor Laboratory, wrote that he was "appalled" by Maddox's comments, arguing that the requirements for industry and academic authors should be the same: "If industry wishes to keep its sequences secret, then it should not publish."[39]

Maddox, however, viewed the matter as "an issue of principle touching the relations between journals ... and their contributors on the one hand and their readers on the other." He seemed especially worried about journals "being turned into instruments of law enforcement," and, foreseeing a slippery slope, he raised the specter of preconditions of publication being extended beyond "the wishes of the databanks" to include compliance with regulations on recombinant DNA, embryo research, animal welfare, and national secrecy.[40] *Nature* continued to "request" that authors deposit sequence (and crystallographic coordinates) with the databases for another six years before it instituted a policy requiring sequences to be deposited with the databases as a condition of publication.[41]

Despite their opposing views, Maddox and the supporters of mandatory submission agreed on one crucial point: that mandating direct submission represented a significant shift in scientific publication, a world poised to be reshaped as new forms of electronic communication were introduced. Maddox's commitment to the notion of a journal of record, fully autonomous and tied to the printed page, contrasted sharply with the views of those who thought editors had "a responsibility to steer the publication process through its next evolutionary stage."[42]

As new knowledge-control regimes take shape, normative questions often arise about the extent to which control relationships should depart from the precedents established by existing regimes. The creators of GenBank quite self-consciously recognized that mandatory direct submission was a new form of scientific communication, which they dubbed "electronic data publishing."[43] Underlying the regime was the distinction between "sequence data" and "results based on those data," coupled with the conviction that results belong in journals, whereas sequence data belongs in databases.[44] Governing this new mode of publication entailed addressing a number of normative questions about what principles and practices to follow.

One prominent issue was the timing of access to sequence data. Under the regime of staff-driven collection, database users could not access a sequence until a paper containing it had appeared in print and, subsequently, staff had entered the data into GenBank. Direct submission removed this constraint, giving GenBank the capacity to make data available online at any time after submission. Allowing access to sequence information before the paper based on it was published, however, would encroach on a traditional privilege in the laboratory regime: to exercise strong control over the distribution of data and materials until after publication of a paper relying on them. Building a direct-submission regime thus entailed settling questions about keeping sequences confidential. (Should GenBank maintain confidentiality, and if so, for how long? Until acceptance of the paper? Until the paper appeared in print? Should authors be able to request additional delay?) GenBank had to actively decide, taking account of established control relationships governing the adjacent regimes of the journal and the laboratory. In this case, GenBank offered to hold submitted data in confidence until a related paper appeared in print—a policy that closely matched the entitlements and burdens inscribed in the traditions of the journal regime.[45]

GenBank also had to decide how to review incoming submissions. Invoking the distinction between "scientific results" and "sequence data," its leaders concluded that scientific results need "peer review and an essentially free-form medium like the printed page," whereas sequence data need "a largely automatic form of quality control and a highly structured, electronic format to be useful."[46] GenBank argued that its "increasingly automatic" checks for errors—combined with the advantages of handling data only electronically (thus eliminating transcription errors, the largest source of errors)—had "actually increased the level of review that data receive before being made public."[47] Annotation was also expected to improve because GenBank made the producers of the data responsible for this task. Rather than having GenBank staff struggle to understand a wide range of organisms and biological systems, this approach "shifts the responsibility for understanding the data to the people who already understand it."[48]

GenBank also faced questions about how to handle refinements or corrections to data after submission. Journals, of course, allow papers later deemed to be incorrect to remain part of the permanent record. Errata and retractions are used sparingly, and editors do not march through the world's libraries tearing out pages containing rejected knowledge claims. But electronic systems make it feasible to "depublish" information by simply deleting content. Should GenBank preserve a record of who submitted what and when? Or should the data constantly be massaged to replace earlier submissions with updated and corrected information? Choosing the latter course would entail conducting authoritative reviews of database contents. Who should be entrusted with this task? Could such reviews be performed in a timely manner? Should scientific institutions create knowledge-control regimes that allow "uncorrected" versions of knowledge to disappear down a virtual "memory hole"?[49] Given the accelerating rate of sequencing, it is no surprise that GenBank eschewed policies requiring reviews of already submitted data.[50] But GenBank sought a way to reconcile the goals of preserving an archaeological record and offering updated information. Its solution was "[not only] to store unchanging reports of the sequences as originally presented but also to provide a syntax by which a complete, correct, and up-to-date picture of the biological reality can be built up out of a composite of these reports."[51]

Regarding academic credit, the GenBank leadership acknowledged that a sequence was a publication only "in a sense," and although those who submitted sequences would be credited with a bibliographic reference in the database, the value of this credit would be small. GenBank defined its task as maintaining records of who had submitted what, not evaluating the significance of contributions. ("The research community"—not the database—"will always decide de facto the relative importance of the generation of data."[52]) The "mere" production of sequence information—absent biological insights derived from analyzing it—would be regarded as the execution of a "routine" "technical" procedure rather than a "scientific" achievement. Generating sequence data would by itself garner little credit.

The regime of mandatory direct submission became a durable settlement. Enforced by the journals and accepted by the many biologists who produced sequence information during their everyday work, it remains in force today. But when large-scale sequencing projects emerged at the outset of the HGP, this regime proved incapable of governing this new domain. Large-scale sequencing became a special case, and the rest of the chapter focuses on the challenges of governance that it raised.

The Second Destabilization: Large-Scale Sequencing and the "When" Question

If during the 1980s generating sequence data was increasingly regarded as little more than a "technical" step along the road to "biological findings," then the large-scale sequencing pilot projects launched at the beginning of the HGP were something of an exception. These very visible projects aimed to sequence larger contiguous strips of sequence than had ever been sequenced before. They also sought to pass major scientific milestones, such as producing the first complete DNA sequence of a single-cell organism, the first eukaryotic genome sequence, or the first sequence of the genome of an animal. Papers reporting such achievements would surely be prominently published in top journals and earn coverage in science journalism and even footnotes in history. Nevertheless, many genome researchers expected that large-scale sequencing projects, setting aside the most dramatic "firsts," would be stigmatized as "service" activities or "technological" successes rather than "creative" "biological" accomplishments. Many of my interlocutors also anticipated that the laurels would go mainly to laboratory heads, not to other members of the research teams. Some of them also predicted that the recognition granted for large-scale sequencing would decline as the process grew more routine because everyone would forget how difficult it once had been to carry out.

The principal goals of the large-scale pilot projects were to experiment with sequencing strategies, generate large quantities of data, and pass some biological milestone. Most large-scale sequencers also wanted to do "a little science" using the data, in which they would conduct a "preliminary analysis" and prepare a substantive publication describing the number and density of genes and other interesting features of the sequence. But because these projects were long-term endeavors, waiting to publish a paper and submit the data until the sequencers had completed their project would prevent other scientists from seeing the data for years. Scientists all over the world wanted to obtain these unprecedented quantities of genomic information as soon as possible. Even if incomplete, the continuous strips of sequence that the large-scale sequencers aimed to produce could be useful for many purposes, such as hunting for genes, studying genome organization, and testing algorithms for sequence analysis. Some members of the broader molecular biology community wondered if large-scale sequencers would use their early access to data to "unfair" advantage. This situation raised what we might call the "when question": *When should large-scale sequencers submit data to the databases?*

The direct-submission regime—in which timing is determined by publication decisions made in the journal regime—failed to provide an answer.

Direct submission offered a workable settlement in the case of "ordinary biology" because scientists typically wanted to publish a paper as soon as they found a biologically significant "result." This logic did not apply to large-scale sequencing projects. The large-scale sequencers could not afford to divert much time to investigating "potentially interesting" biological phenomena unrelated to the technological challenges of sequencing. Papers on "biology" therefore could not be counted on to trigger submission. In principle, large-scale sequencers could choose to publish papers and submit data whenever they reached specific production milestones. However, many plausible milestones existed, including completing the sequence of the entire target organism; completing 25 percent of the sequence; completing a chromosome; or completing a cosmid (approximately 40,000 base pairs). Alternatively, submission might follow a fixed schedule (e.g., partial results would be submitted every three months) or be triggered by a clock that started when a strip of sequence was generated.

The "when" question raised the normative problem of allocating entitlements and burdens among several agents, most notably the laboratories generating sequences and the rest of the biological research community. The NCHGR and DOE initially had no official policy, although clearly their general orientation supported releasing data with minimal delay. Some pilot projects also stressed their commitment to speedy release of the data. The scientists running the C. elegans sequencing project were especially strong advocates of rapid release. The community of scientists who studied the worm bore structural resemblance to the group of early Drosophila geneticists described by Robert Kohler: they were a small and rather newly formed group, bound together by a strong sense of being part of a collective endeavor, with dense social networks and a "moral economy" emphasizing cooperation.[53] The leaders of the worm project, John Sulston and Robert Waterston, quite self-consciously hoped to set a precedent for "rapid release" applicable not only to the worm but also to other organisms, notably the human. They envisioned a sequence being released weeks or at most months after it was generated.[54]

To some extent, the large-scale sequencers' conceptions of what was fair differed, with position in the laboratory hierarchy related to the perception of what was at stake. Laboratory heads, because their scientific identities were well established, mainly stood to benefit from submitting data quickly, which allowed them to visibly demonstrate productivity and publicly display generosity. For them, the timing of submission raised internal managerial and moral concerns about developing the careers and maintaining the morale of the younger scientists in their laboratories. At the other end of the hierarchy, the issue was unimportant for most technicians because few of them imagined themselves building careers in genome research.[55] However, for younger PhDs in HGP laboratories, questions about personal credit were acutely felt matters; they needed to make distinctive personal contributions in order to produce a viable scientific identity. They expected that those whose work focused exclusively on sequencing "anonymous DNA"—a term genome scientists use to describe arbitrary DNA fragments— would themselves remain anonymous. Some sought credit for developing sequencing strategies and technology, but others wanted sufficient time to do a career-enhancing analysis of the data.

In some large-scale sequencing projects, providing rapid access to data thus stood in tension with the exigencies of scientific careers and the prevailing concepts of reward for work in molecular biology. During a presentation at a small workshop in Washington in early 1992, for example, an American scientist, BRAD, expressed resentment about pharmaceutical companies badgering sequencing groups to release sequence information shortly after its production. I later asked him to elaborate in a private conversation:

Brad said that it was outrageous for a company or anyone else to think that we are obligated to turn the data over to them right away. We want to do "some science" on it first and give the people who spent a long time producing it a crack at exploiting it scientifically a bit before it is submitted. Brad also said that you couldn't wait forever to submit it. He saw questions about the right ethics in this case as a real issue, and he thought that the young people who were "busting their butts" to produce sequence data deserved a chance to get a publication out of it. He said that a grant review committee told the BLUE LAB: "This is a great proposal, but we don't want you to do any science on the data, just turn it over." He thought this was fundamentally unfair.[56]

For Brad, the question of when sequence data should be released was entangled in issues about what is enough data and what is enough analysis to underwrite a paper of enough interest to be career enhancing. In a world where generating the data was increasingly unlikely to impress people, the quality of the analysis would grow more important. Although Brad supported some delay, he expected it to be fairly short because he anticipated that the tools for computational analysis of sequences would grow increasingly powerful, allowing more analysis to be accomplished in less time.

At the outset of the HGP, the pilot sequencing projects developed their own laboratory policies about the timing of database submissions. The RED LABORATORY initially decided to submit data whenever it completed a cosmid (about 40,000 bases). In the PURPLE LABORATORY, the sequencing strategy called for dividing the genome into parts about 100,000 base pairs long and sequencing them one at a time. This strategy was intended to simplify the problem of closing the last gaps, but the Purple Laboratory also planned to publish a paper and release the data each time that it finished one of these 100,000-base-pair fragments. These scientists conceived of the laboratory as a "sequencing facility," and its role was to develop technology, generate sequences, and move on to the next organism, not to "play" with the biology forever. They stressed that any delay between completing a fragment and releasing it would be short, defined by the period required to do a "preliminary analysis" and prepare a paper describing the sequence, identifying protein coding regions (which would be putative genes), and reporting on their number and spatial distribution. However, the Purple Lab planned to leave most of what they called the "biology" to other scientists.

Regime 3: The Six-Month Rule (1992 to 1996)

Because these large-scale sequencing projects developed their own policies on the timing of data submission and publication, it is not surprising that some scientists expected or perceived strategically motivated delays. Worries about strategic delays—in large-scale sequencing and mapping— provoked calls for a standard rule aimed at preventing genome project laboratories from engaging in such practices. In the United States, the idea that gained traction during 1991 and 1992 was a rule specifying that data and materials created using genome program funds should be "made public" within six months of being generated. In informal settings, my interlocutors discussed several other possibilities—30 days, three months, a year—but six months emerged as a "reasonable" delay. After a group of science advisers recommended that the NIH and DOE prepare a joint statement on this issue, the agencies drafted a document, and it was unanimously approved at the NIH–DOE advisory committee meeting in December 1992. The document stated that "consensus is developing around the concept that a 6 month period from the time the data or materials are generated to the time they are made available publicly is a reasonable maximum in almost all cases. More rapid sharing is encouraged. ... Whenever possible, data should be deposited in public databases and materials in public repositories." To put teeth in the statement, the document stressed that applicants for genome project grants would be required to describe their plans for making data and materials available and that these plans would be reviewed "to assure they are reasonable and in conformity with program philosophy."[57]

Many scientists, including those present at the meeting where the policy was approved (which I attended), recognized that policing the six-month rule might not be straightforward. People outside the laboratory could not ascertain when any particular stretch of sequence had been generated. Moreover, the event marking the moment of data "generation" and starting the six-month clock was subject to interpretive flexibility. Should each "read" trigger its own clock the instant that it came off a sequencing machine, or should the clock start only once some larger contig had been assembled? In principle, many milestones could be chosen to trigger the clock, ranging from the single read to the cosmid to the chromosome to the entire genome.

Further complications arose from the ambiguous definition of the term *finished sequence*, which had not stabilized. Genome scientists typically considered the sequence of a genomic region to be finished if it contained no "unavoidable" gaps and had an "acceptable" level of sequencing errors. Most gaps in a stretch of sequence could be closed, often through painstaking work.[58] There was also the issue of accuracy—that is, the percentage of miscalled bases (e.g., the sequence ACTTG reported as A*G*TTG), inserted bases (e.g., the sequence ACTTG reported as ACTT*T*G), or deleted bases (e.g., the sequence ACT*T*G reported as ACTG).[59] Repeated sequencing of the same DNA could reduce the number of errors, but a standard definition of accuracy had not been developed in the early 1990s, and there was some debate about how accurate data needed to be. Moreover, Brad, among others, pointed out that "checking" data prior to starting the clock and doing "science" on the data afterward might look similar, at least from outside the laboratory.

The HGP leadership in effect ignored these ambiguities. As often occurs when framing policy, it seemed more important to establish *a* rule than to fret too much about how an ambiguous rule might later be interpreted. As events unfolded, the potential difficulty of policing the six-month rule did not prove to be a huge problem for the NCHGR. Only in a few cases did it have to "really lean on people" to enforce its data submission rules.[60] During the early 1990s, the successful large-scale sequencing groups tended to submit sequences in a manner that most observers considered relatively timely. Seeking to demonstrate productivity and stay in the game, the large-scale sequencing groups wanted to submit sequence information to the databases because the quantity of data submitted was the most visible metric of laboratory output. Moreover, the collaboration led by John Sulston and Robert Waterston that had begun with *C. elegans* and expanded to human sequencing was strongly committed to submitting data rapidly. Sulston and Waterston's influence was enhanced by the fact that theirs was the most successful of the early sequencing projects. Nevertheless, there was variation. For example, the policy of the European yeast-sequencing program, a coordinated network of small laboratories, inspired criticism, especially in the United States. This program treated the yeast chromosome as the unit of submission, and it explicitly allowed participating laboratories to hold sequence data as their "property" to be mined for publications and (at least in principle) patents until publication of a paper on an entire chromosome. The leaders of the European yeast project justified their approach on the grounds that the scientists who had worked hard to generate the sequence deserved to reap some rewards.[61]

As the pilot sequencing projects began reaching milestones, such as completing a yeast chromosome, they began publishing broad "overview" papers along the lines of what the large-scale sequencers had anticipated a few years earlier.[62] The typical overview paper described what had been sequenced, outlined the methods used, and provided a "preliminary analysis" of biological features such as the number and density of genes. For example, when the European yeast-sequencing network completed yeast chromosome XI, it published a paper in *Nature* in 1994 describing the chromosome; that paper became the key publication documenting the contribution of 108 authors who had participated in the project.[63] Similarly, when the entire sequence of the yeast genome was completed in 1996, the collaborating groups published an overview paper.[64] Such overview papers became the normal way to mark the completion of a project and to credit sequence producers.

The Third Destabilization: Anticipated Coordination Problems

The HGP began to seriously consider shifting its focus to sequencing in 1994 and 1995 (discussed further in chapter 7). John Sulston, director of the Sanger Centre, and Robert Waterston, head of the genome center at Washington University, pressed hard for a transition from mapping to sequencing. They also argued that major sequencing centers needed to coordinate work and ensure speedy data submission and publication. Sequencing remained labor intensive, and some 99 percent of the human genome awaited sequencing, so HGP scientists and funders had every reason to avoid duplication of effort. A free-for-all, with laboratories sequencing overlapping regions, whether inadvertently or in competitive races, seemed extremely unattractive. The need to address coordination and competition among the large-scale sequencers loomed large, and controlling the politics of timing came to seem more pressing. In this context, the six-month rule began to look inadequate.

Regime 4: The Bermuda Principles (1996 to 2003)

Sulston and Waterston joined forces with Michael Morgan of the Wellcome Charitable Trust, funder of the Sanger Centre, and the three convened an international meeting of the main sequencing laboratories. Their goal, as Sulston later described it, was "to hammer out a strategy for deciding who would do what, and how the data would be managed."[65] In light of the struggle to control ESTs, Sulston and Waterston also worried about the possibility that commercial firms might undertake large-scale sequencing of the human genome in order to build a portfolio of patents.[66] The meeting took place in February 1996 in Bermuda, a site selected because it was "neutral" ground; a British territory near the United States seemed appropriate given that the US government and the United Kingdom's Wellcome Trust were the only entities committed to spending the billion-dollar-level sums needed to sequence the human genome.[67] What other countries would end up contributing remained unclear, but the organizers invited representatives of large-scale sequencing groups from the European continent and Japan as well as funding-agency officials.

In Bermuda, the sequencing groups agreed to avoid duplication of effort by dividing up the human genome, allowing each group to claim regions that they planned to sequence. Most importantly, the meeting organizers proposed that the HGP should require large-scale sequencing laboratories to rapidly submit sequence information to the databases. To Sulston's surprise, the attendees reached an unprecedented agreement mandating that sequence information must be submitted to the databases *immediately* after it was generated.[68] The meeting thus established the principle that the HGP's large-scale sequencing laboratories would practice rapid prepublication release of sequence data. The "consensus" underlying the agreement proved to be thin, but it enjoyed the support of most of the largest sequencing laboratories and the largest funders—namely, the Wellcome Trust and the US NHGRI and DOE.[69] The Bermuda Principles, as they came to be known, imposed extremely stringent data-release requirements on large-scale human sequencing centers. Newly generated, "preliminary" sequence would be released as soon as possible, with some centers submitting assemblies greater than 1,000 bases—a threshold reached by joining just a few overlapping reads—on a *daily* basis.[70] Laboratories had to submit "finished" sequence data immediately.[71] In effect, the Bermuda Principles required sequence information to be submitted within 24 hours of its generation.[72]

This unprecedented policy served not only to make data available but also to enhance transparency by making the *fact* of its submission visible. Beyond speeding scientific progress, the Bermuda Principles also served as a means to demonstrate that the large sequencing centers—which were expected before long to be spending tens or hundreds of millions of dollars per annum—were behaving responsibly. "The people who were going to be given this franchise to sequence large parts of the genome … didn't want to be subject to the criticism that they had a leg up and were taking unfair advantage" of their position to acquire publications or patents, one NHGRI official explained.[73]

In the aftermath of the meeting, the Bermuda Principles remained controversial. One issue was quality control. Daily releases would consist of "unfinished data" based on the output of single runs of the automated sequencing machines, which software would assemble into small contigs. The information would often contain errors, and transforming it into "finished" sequence—much longer contigs with a low error rate—could take months. Advocates of the Bermuda Principles contended that even unfinished data were of immediate scientific value. Critics, such as Craig Venter and Mark Adams, both of Venter's nonprofit institute TIGR, argued that low-quality data could cause confusion.[74] To such worries, advocates of immediate submission replied that "the unfinished sequence is not a substitute for the finished product but constitutes a transient, dynamic buffer of finite size." Sequencing centers thus would be responsible for ensuring that data do "not languish in the unfinished category."[75]

Incorporating the Bermuda Principles into the regime governing the HGP thus imposed two new duties on large genome centers: to release unfinished sequencing information immediately and to replace it with finished data reasonably quickly. Implementing these modifications required help from the databases. GenBank, which had originally been reluctant to accept unfinished, rapid-release data owing to concerns about quality, decided to cordon off the stream of incoming large-scale sequencing data, establishing a special part of the database known as the High Throughput Genomic Sequence Division (HTG Division). Large-scale sequencing laboratories could submit preliminary, rapid-release data directly to this part of the database. A prominent warning accompanied each unfinished sequence in the HTG Division, as this example illustrates:

WARNING: Phase I High Throughput Genomic Sequence

This sequence is unfinished. It consists of 6 contigs for which the order is not known; their order in this record is arbitrary. ... When sequencing is complete, the sequence data presented in this record will be replaced by a single finished sequence with the same accession number.[76]

When the submitting laboratory replaces a preliminary HTG sequence with a "finished" version, the warning is removed.

Quality control was not the only point of contention about the Bermuda Principles. A second issue concerned friction between the Bermuda regime and the regime of publication in journals. Adams and Venter, for example, worried that if papers were sent to journals long after the underlying sequence was fully available on the Internet, peer reviewers might decide that "the novelty of a manuscript has been compromised." They also argued that rapid prepublication release showed "indifference" toward the scientist's intellectual effort: "Scientific custom has held that the scientist should be allowed to communicate to the research community what was achieved and how it was done, to analyze and comment, not only so that careful critical evaluation can be made, but also out of respect for the researcher and the achievement."[77] Adams and Venter thus equated the large-scale sequencing center to the autonomous scientist, a figure whose identity and privileges they sought to defend. To put the matter in Hohfeldian terms, they objected to replacing the privilege to decide whether and when to transfer holdings (and to whom) with the duty of daily public release.

The Bermuda Principles established an important transformation in the control relationships governing interaction among genome centers, databases, and funding agencies. The largest funders—the Wellcome Trust and the US agencies—and the best-funded, largest sequencing centers (with the exception of TIGR) supported the agreement. Countries making smaller financial commitments followed suit.[78] The US genome program, building on its established regime that defined genome centers as producers of data for community use, made its requirements more stringent by eliminating discretion about timing. The Sanger Centre and Washington University watched their preferred policy of rapid release be adopted not only by the United States but also by other national programs. The Bermuda Principles, which were reaffirmed in 1997 and 1998, thus gave new concreteness to the HGP's international aspirations.[79] HGP sequencing centers followed the policy of rapid release until the project's official endpoint in 2003. Even during the competition with Celera Genomics (discussed in chapter 7), the HGP continued to post data on a daily basis.

The Fourth Destabilization: Disputes over Legitimate Use

The Bermuda Principles remained in place for the duration of the HGP. But this regime also set the stage for events that destabilized the prevailing mode of awarding credit to large-scale sequencers. Rapid prepublication release constituted a new category of data that did not fit neatly into the familiar ontology of the publication regime (e.g., published versus unpublished)—namely, sequence information that was "available" in the databases but "unpublished" in the peer-reviewed literature. To describe these data in precise and relatively neutral terms, I use the neologism *UJAD data*, short for data "unpublished in journals and available in databases." The liminal status of UJAD data set the stage for a bitter dispute in the early 2000s about who could use them for what purposes and when.

During the 1990s, publishing a "primary" "overview" paper had become what genome scientists viewed as the "traditional" way (in this novel and fast-moving field) for large-scale sequencers to get credit for their work. As is often the case with knowledge-control regimes, this expectation was embedded in ongoing practices, not grounded in codified rules. However, in several instances, large-scale sequencers were shocked to find that other research groups had downloaded the UJAD sequence information from a nearly finished project, analyzed it using bioinformatics tools, and published overview papers that preempted similar overview publications by the group that had produced the data.

One prominent case involved a collaboration spearheaded by Lee Rowen of the University of Washington. The collaborators aimed to sequence the DNA underlying a component of the immune system known as the major histocompatibility complex (MHC).[80] The plan was to sequence the MHC region in two closely related species, the human and the mouse, and prepare a major paper analyzing the similarities and differences.[81] In keeping with the Bermuda Principles, Rowen and her collaborators submitted unfinished sequences to GenBank as the sequencing progressed, providing updates and annotations as segments of the MHC were finished. By 1999, they had completed much but not all of the sequencing. As Rowen and her colleagues worked to wrap up the project, they learned that two separate research groups, one from Australia and one from the United States, had downloaded all of the data from the MHC project and submitted "overview" papers to *Immunology Today*—papers that closely resembled the publication with which Rowen and her collaborators were planning to mark the completion of their project. A peer reviewer had raised a red flag, suggesting that the data and annotations might have been improperly appropriated. The journal informed Rowen and her collaborators, who vehemently objected, arguing that they had a "right" to publish an overview paper about the data that they had worked so hard to produce. They also refused an offer of coauthorship, rejecting what they saw as forced collaboration and arguing that the Australian and US groups had no right to use the data in this way. In July 2000, *Immunology Today* published the "overview" papers despite these objections, and the dispute, which had begun in a series of emails, went public.[82] Rowen and several other large-scale sequencers asserted what some called a "right of first publication" that entitled them to author an "initial paper" giving an "overview" of the data.

The ensuing controversy divided the genomics vanguard. The question was often framed as a dispute between two types of agents: "sequence producers" and "sequence users." In fact, it was not a simple, two-sided conflict between producers and users (or, for that matter, between sequencers and bioinformatics specialists). None of these types of scientists was completely unified. Many genome scientists considered it extremely inappropriate for a "sequence user" to download UJAD data and rush out an overview paper before the laboratory that had generated the sequence published an analysis of its "own" data. They regarded the assemblage of UJAD data from a nearly completed project as akin to a "personal communication," appropriate to use only with permission. Some scientists, such as Stanford's Richard W. Hyman, argued that misappropriating these data constituted a form of scientific misconduct similar to plagiarism.[83]

Others placed UJAD data in the category of "published information." All of the information in GenBank had long been considered fair game for anyone to analyze in any way that they chose. To some, this precedent seemed completely applicable to the MHC sequence case. Indeed, some database managers and computational biologists thought that even limited encumbrances on use conflicted with the very idea of a "public" database, which they envisioned—in keeping with the Internet imaginary of "information at your fingertips"—as a vast repository from which anyone could selectively extract sequences, combining and analyzing them freely, without worrying about their provenance.

For those who argued that provenance could not be ignored, the challenge was determining precisely which uses required the sequence producer's permission.[84] The NHGRI tried to draw that line in a December 2000 update to its data-release policy: "NHGRI believes that a reasonable approach is to recognize the opportunity and responsibility for sequence producers to publish the sequence assembly and large-scale analyses, while not restricting the opportunities of other scientists to use the data freely as the basis for publication of all other analyses, e.g. of individual genes, gene families and other projects at a more limited scale."[85]

Defenders of the idea that sequence producers retained no residual rights sometimes suggested that large-scale sequencers were being funded to produce and submit information, not to analyze it. Sequencers responded that they were running sequencing projects "for a reason"—namely, that they were "scientists" and as such were naturally interested in analyzing the data that they produced.[86] The large-scale sequencers had an "intellectual stake" in their work and had "invested their careers" in the area and so should not be treated simply as "benefactors" who provided resources to others without compensation in the "currency of the realm": publication. My interlocutors also argued that if the scientific community failed to protect sequencers from being "scooped," then the HGP's commitment to free and open data would become "a one-time thing."[87]

Regime 5: The Fort Lauderdale Agreement (2003)

The stakes in the UJAD debate extended beyond the immediate parties and beyond the genome project. As the HGP drew to a close from 2000 to 2003, the NHGRI began considering other major sequencing projects, and many members of the genomics vanguard wanted rapid release to continue. Moreover, after Celera was founded in 1998, the possibility that the collapse of rapid release would lead DNA sequence data to be privatized worried many HGP scientists, strengthening their desire to settle the UJAD issue. In 2001, JIM, an official with the NHGRI, outlined for me a possible "compromise." The idea, which was discussed at Cold Spring Harbor in May 2001, was for large-scale sequencers "to be up front and clear about what they're intending to do" in a single, initial publication presenting the "data set."[88] This vision rested on the "traditional" concept of using the "overview" paper to award credit for sequencing. Once that initial paper had appeared, then the sequencing laboratory's claims would vanish, and the data would become available for anyone to use for any purpose. In effect, large-scale sequencers would be formally granted the limited and temporary right of first publication that Rowen and her colleagues had claimed.

Implementing such a system, however, posed practical questions, as RANDY, a journal editor who liked the principle of registering claims, suggested:

GenBank should have a website where you register a project, and you say, "I'm sequencing the aardvark and my intention is to ... publish the following analyses. ... You're welcome to look at my data, but you cannot use all of my data for doing that [same] kind of analyses until I've published my paper. That paper will be published in three months or eight months or whatever." ... It would be ambiguous. That's a slippery slope, too, because who polices that? What happens if I claim the world? What happens if I take too long? Who mediates that? ... Some of the people who were involved in the discussion earlier this week said, "Well, yeah, you may start out today thinking I want to do ABCD," and then a month later when you're halfway through generating data, you say, "Well, I also want to do FGHI." Do you have a right to keep updating that?[89]

Beyond the difficulties of defining the substantive and temporal scope of entitlements and empowering some authority to register claims, police violations, and settle disputes when they arose, the implementation of such a system would entail an unprecedented change: allowing scientific authors to claim exclusive rights to publish in the future about results that they had yet to achieve.[90] To be sure, the relevant claims would specify accomplishments that were likely to be realized, such as applying standard analytic techniques to a particular sequence. But recognizing such rights would entail extending authorial rights beyond *consummated* achievements to include claims on *anticipated* ones—a fundamental change in the logic of the publication regime.

In January 2003, the Wellcome Trust sponsored a meeting in Fort Lauderdale, Florida, that aimed to find an appropriate solution. The meeting brought together about 40 people, including representatives of the sequence databases, the large-scale sequencing groups, journal editors, computational biologists, and others interested in large-scale biology. Their discussions produced an 1,800-word document, soon dubbed the "Fort Lauderdale Agreement." It reaffirmed the Bermuda Principles and argued for extending the practice of immediate data release beyond the HGP. The document outlined the governing frame of a knowledge-control regime designed to address the problem of prepublication use of UJAD data.

Vanguard visions that depart too much from prevailing practices typically face obstacles in achieving acceptance. In this case, the challenge of finding a "compromise" not only entailed balancing the claims of the various parties but also doing so without advocating changes that would greatly disrupt the governing frames of the journal, laboratory, and rapid release regimes. The Fort Lauderdale Agreement accomplished this by advancing informal restrictions on prepublication use and simultaneously establishing limits on those restrictions in three ways.

First, the proposed regime had a limited jurisdiction. It pertained only to a narrowly bounded set of research endeavors—namely, "community-resource projects" (CRPs): large-scale efforts "specifically devised and implemented to create a set of data, reagents or other material whose primary utility will be as a resource for the broad scientific community."[91] In effect, the CRP concept built on the governing frame of the HGP, which defined the "genome program" as sufficiently different from "ordinary biology" to necessitate that it be governed in different ways. The Fort Lauderdale Agreement generalized this distinction to include future "infrastructure" projects in genomics, such as the HapMap Project, and in emerging areas such as proteomics.[92]

Second, the proposed regime was based not on formally codified "rights" but on a more informal (although still lawlike) system grounded in mutual respect and self-restraint. Eschewing the language of *rights* (a word that does not appear in the document), the Fort Lauderdale Agreement envisioned a regime grounded in "normal standards of scientific *etiquette.*" It was built around an ontology that implicated three types of agents— "funding agencies," "resource producers," and "resource users"—in "a system of tripartite responsibility" for governing "community resource projects."[93] It proclaimed that

• *Funding agencies* should require CRPs to institute "free and unrestricted data release" as "a condition of funding." They should also support "the ability of the production centres to analyse and publish their own data."
• *Resource producers* should make data "immediately and freely available without restriction." They should also "recognize that even if the resource is occasionally used in ways that violate normal standards of scientific etiquette, this is a necessary risk" to achieve the benefits of immediate data release.
• *Resource users* should cite and acknowledge the resource producer's work and "recognize that the resource producers have a legitimate interest in publishing prominent peer-reviewed reports describing and analyzing the resource that they have produced."[94]

The agreement normalized this particular configuration of responsibilities by positioning it as grounded in well-established standards of scientific conduct.

Third, the proposed regime aimed to delimit the claims of the resource producers, whom it encouraged to publish a "Project Description" at the outset of each project. The Project Description would be "a new type of scientific publication" that would inform the scientific community about the details of the project, provide a timeline for data production and release, and describe the scope of the analyses that the resource producer intended to undertake. Resource users, in turn, were expected to "respect the producer's legitimate interests … while being free to use the data in any creative way."[95] The agreement thus aimed to constitute two categories of uses: those that legitimately belonged to the resource producers and were specified in the Project Description and those that anyone could pursue. Mutual respect was to guide action: "There should be no restrictions on the use of the data, but the best interests of the community are served when all act responsibly to promote the highest standards of respect for the scientific contribution of others."[96]

The NHGRI soon made the system outlined in the Fort Lauderdale Agreement its official doctrine. Backed by the authority of funding agencies—as well as by the "etiquette" of mutual respect—the agreement thus constituted a knowledge-control regime intended to allocate appropriately the entitlements and burdens surrounding UJAD data.[97] The principle that large-scale "infrastructure" projects should ensure rapid and widespread availability of data became a recognized norm of what some observers referred to as the "postgenomic" era.[98]

Conclusion

Transformative scientific change may be accompanied by a dynamic of stabilization and destabilization that produces a cascade of distinct knowledge-control regimes. The rise of sequence databases and large-scale sequencing is widely understood as a scientific revolution, and, as I have shown in this chapter, it provoked ongoing reordering of knowledge-control regimes. Change began in the narrow world of the first sequence databases, such as Goad's project at Los Alamos, but it soon spread to encompass adjacent jurisdictions. Under the first regime examined here—staff-driven collecting—separate regimes governed GenBank and the journals, which operated as distinct jurisdictions without direct, cooperative interactions. GenBank was wholly dependent on journals for information and as an epistemic gatekeeper. Editors, peer reviewers, and the authors of manuscripts controlled the flow of sequences into journals, which in turn defined what sequences would be available to GenBank and when. Normative justification for GenBank's use of sequence data rested on the moral order of the journal regime, especially its conception of the entitlements of readers of published work. The control relationships underlying the journal regime and the laboratory regime remained untouched.

Soon, however, the explosion of sequence data and the growing scientific importance of the databases destabilized the regime of staff-driven collecting. Mere coexistence of journals and databases—as adjacent jurisdictions—no longer seemed sufficient. Negotiations involving the databases, the journals, and the funding agencies established the second regime, mandatory direct submission. This regime imposed new duties on agents in the journal regime, requiring authors to submit sequence data and provide annotations. Decisions in the journal regime continued to control matters of timing because acceptance of a paper triggered database submission. However, gatekeeping was institutionally bifurcated as the distinction between "results" and "data" grew sharper and was inscribed into institutional roles: journals managed peer review of "results," and databases enhanced quality control of "data" with automated checks. Normative justification for direct submission remained grounded in the principles of the journal regime, especially the rule that data underlying published papers should be made available. Meanwhile, the privilege of laboratories to decide when to submit papers remained untouched.

The launch of the HGP and other programs of large-scale sequencing raised problems of governance that the direct-submission regime could not address. Many actors worried that allowing laboratories and journals to control the timing of submission to databases would produce unacceptable delays. The normal regime of the laboratory seemed to grant too much autonomy to the large-scale sequencer. New regimes building on the conception of the HGP as different from "ordinary biology" took shape, with funding agencies, genome laboratories, and databases negotiating about regime structure. The third regime, the US six-month rule, defined the large-scale sequencing laboratory as a particular type of facility—one that could justifiably be subjected to special obligations to submit data. This regime curtailed the autonomy of large-scale sequencing laboratories, imposing new duties on them. Within a few years, however, the anticipated transition to sequencing destabilized the six-month rule, which genome project leaders deemed insufficient to manage the challenge of coordinating a concerted effort to sequence the human genome once it scaled up.

The fourth regime, the Bermuda Principles, further curtailed the large-scale sequencer's privileges, making sequence information available nearly instantly and completely detaching the timing of database submission from the temporal rhythm of journal publication. The regime also enhanced transparency, with the goals of easing coordination problems and reducing concern about the "hoarding" of data. Normative justification rested on claims about the obligations of large-scale sequencers to collective interests in data access. However, the Bermuda Principles created new instabilities. With large-scale sequencers putting "preliminary" sequence data into widespread circulation, questions about the control relationships pertaining to this information grew salient. How did this information—which was formally "unpublished" but available to all—fit into the categories of the regime of scientific publication? Could anyone use it for any purpose? Ambiguous control relationships set the stage for bitter disputes about the terms of use.

The fifth regime, the Fort Lauderdale Agreement, sought to manage these disputes by building on the conception of the HGP as different from "ordinary biology" to develop the general concept of a community-resource project—a distinct type of science that should be governed by special rules (including rapid data release). CRPs were envisioned as involving three main types of agents—funding agencies, resource producers, and resource users—who were encouraged to exercise responsibility and restraint.

The rise of sequence databases and large-scale sequencing thus produced ongoing instability in knowledge-control regimes. At first glance, the process of interaction—which produced five regime changes in approximately two decades—might look like chaotic turbulence. But closer inspection reveals a recurring tendency for the settlements reached to tread lightly on the jurisdictions of the journal and the laboratory. Staff-based collecting did not encroach on the journal regime. Mandatory direct submission intruded into the journal's jurisdiction, imposing new duties on authors and journals, but did not impinge on the most important control relationships. Later, when conflict arose about UJAD data, the settlement achieved in Fort Lauderdale carefully limited the authorial claims that sequence producers could request on anticipated results. Etiquette prevailed over rights, and enforcement depended on a loosely defined system of tripartite responsibility. Adjustments to the laboratory regime also took place as unprecedented data-release requirements were imposed on large-scale sequencers. However, these changes applied only to the special case of HGP-funded sequencing facilities, whose work was considered distinct from "ordinary biology"—an ontology of scientific work that the concept of a CRP later generalized to constitute a new but limited category of scientific endeavors. In short, although this period involved ongoing adjustments at points of jurisdictional contact, changes in the well-established knowledge-control regimes of the journal and laboratory were both incremental and bounded in ways that preserved the most important features of their governing frames.

7   Shaping News and Making History

Political revolutions are fought through conflict not only on the ground but also in the arenas of public communication as competing groups construct narratives and invoke imaginaries intended to stir passions, rally supporters, and win hearts and minds. Like their more obviously political counterparts, scientific revolutions also involve struggles on the public stage. For sociotechnical vanguards to advance their visions, they must inspire support and mobilize resources, and when competing factions take shape around divergent visions, struggles to control the direction of sociotechnical change often entail efforts to promote compelling narratives and justifications.

The genomics revolution is a case in point, and the highly visible competition between the HGP and Celera Genomics provides an excellent illustration of a struggle between two competing factions of the genomics vanguard. In May 1998, Craig Venter, the controversial genome scientist of EST fame, and the Perkin-Elmer Corporation, manufacturer of the sequencing machines used in many HGP laboratories, announced that they were launching a private company to sequence the human genome before the HGP could do so. The new company, eventually named Celera Genomics, claimed that it would achieve this goal in half the time for one-tenth the cost, and the HGP and Celera were soon engaged in intense competition. The contested issues were numerous, but the most important questions concerned knowledge-control regimes. What control relationships would govern access to the sequence of the human genome? Would Celera come to dominate the provision of genome data for years to come? More broadly, as the genomics revolution progressed, in what accountability structures would the agents that produce and control genomic information be embedded? Was the sequence of the human genome the common heritage of humanity or something that a company could appropriately own?

This chapter examines the competition between the HGP and Celera, focusing on the struggle to shape how events would be represented to important audiences, not least members of the US Congress and investors. How would the story be emplotted and in what moral terms? How would the protagonists and their characters be presented? And how would their actions—and the knowledge-control regimes that they sought to emplace—be framed for various audiences? As we will see, Celera drew on discourses of government inefficiency and entrepreneurial prowess to frame events and justify a knowledge-control regime intended to make the company into the leading supplier of genomic information for years to come. In contrast, the HGP invoked imaginaries of open science and the public domain to attack Celera's effort to "capture" the genome. These competing narratives of justification were not an afterthought in a narrowly scientific race to produce data but an inextricable and central part of the competition played out on the public stage of mass media.[1]

The chapter begins by describing the state of play in the HGP prior to the formation of Celera and then turns to the period immediately following the announcement of the new company. The creation of Celera represented a serious challenge to the HGP, and I examine how its formation provoked a major response. The company's existence was made public just a few days before the annual Cold Spring Harbor meeting in 1998, where I was able to do fieldwork and interview HGP scientists in the immediate aftermath of this news. After examining the struggle to shape media coverage and how each side framed the controversy, the chapter turns to the making of history through celebratory events, such as President Bill Clinton and Prime Minister Tony Blair's joint news conference announcing the completion of the human genome sequence in June 2000. In analyzing narratives and imaginaries, I pay particular attention to the United States.

From Mapping to Sequencing

Several years before the competition with Celera began in 1998, the HGP had shifted its focus from mapping to sequencing. In 1988, the NRC Committee's recommendation that the genome project proceed in two stages—beginning with a mapping phase and shifting to full-scale sequencing only after improvements in technology—had left open the question of when the sequencing phase should begin. In 1994, the HGP leadership began serious debate about the prospects for orchestrating what was soon being called a "transition to sequencing." During the end of the 1980s and first years of the HGP, most of my interlocutors had told me that the genome would be sequenced not using gel electrophoresis but with one of the "promising" but undemonstrated technological concepts, such as sequencing by hybridization.[2] But in 1993 and 1994, researchers concluded that none of the novel technologies would be developed in time to meet the 2005 target date.[3] The state of the mapping program also raised strategic questions. Genetic linkage maps of the human genome had been completed, as had several initial physical maps—although the physical maps had problems and were deemed inadequate to serve as a starting point for sequencing. This situation raised the question of how additional physical mapping could best lay the groundwork for large-scale sequencing, perhaps by creating "sequence-ready maps" or—as things eventually turned out—by declaring the mapping stage to be "done" and integrating additional mapping with sequencing.

Genome scientists were also growing more confident in the capacity of sequencing based on electrophoresis. In September 1994, Robert Waterston and John Sulston, whose collaboration had become the most productive of the early sequencing projects, began analyzing the prospects for an aggressive transition to full-scale sequencing. They quickly concluded that electrophoresis-based sequencing machines were within striking distance of being able to sequence the genome. To be sure, many incremental improvements would be needed, but Waterston and Sulston concluded that the HGP should aim to complete the human sequence in the year 2000, five years ahead of schedule. Waterston presented their vision to a meeting of US genome center directors called by NIH program officer Mark Guyer in December 1994, and over the coming months Sulston and Waterston presented the idea at a series of genome meetings.[4] By no means did they claim that completing the sequence would be easy, but they argued that it would be done using electrophoresis and that it was time to get on with the job.[5]

Precisely how quickly the HGP should make a transition to sequencing remained a matter of debate, with one of the key issues being how much to concentrate resources in a few centers.[6] The NCHGR, although supporting a transition to sequencing, did not commit to the Waterston and Sulston's radically accelerated timetable. In March 1995, it issued an RFA for "pilot projects for sequencing the human genome."[7] The program was supposed to last for three years, with the idea of gaining technological and organizational experience.

Meanwhile, sequencers were busy reevaluating the bottlenecks—or "reverse salients" in Thomas Hughes's terminology[8]—that limited throughput. One influential genome scientist, OLLIE, explained that he, like many others, had initially focused too much on automating mechanical steps, such as pipetting, when in fact most routine mechanical operations could be done more cheaply and flexibly by "a workforce not all that different than the one that, say, the fast food industry employs." The true rate-limiting step turned out to be the large number of "finishers"—staff members trained to clean up errors and ambiguities in the output of the sequencing machines. These "finishers" spent lots of time manually editing the "reads" generated by the machines into longer, continuous strips of sequence and "trying to decide whether to repeat experiments." Scaling up this approach to complete the genome would require "just thousands of these finishers," said Ollie, which was "totally implausible." TED, a mathematician, gradually persuaded Ollie that most of the finishers could be replaced with computer programs. Algorithm and software development could make a huge dent in the finishing problem.[9]

A variety of strategic considerations also motivated the advocates of shifting resources to sequencing. In 1994 and 1995, some genome project leaders continued to worry that people would conclude that completing the human sequence was unnecessary in light of the development of ESTs (discussed in chapter 5).[10] In addition, HGP scientists and NCHGR officials worried that if progress sequencing was too slow, people might lose interest in completing the human sequence before a quality product was delivered. Some HGP leaders also worried that sequencing technology had reached a point where a private company might launch a major sequencing project with the goal of capturing control of sequence information, a worry raised in part because the rise of HGS and Incyte suggested that entrepreneurs might be able to construct viable business models. As early as May 1994, a confidential memorandum circulated within the NCHGR warning that if the HGP did not effect a transition to full-scale sequencing soon, "privately funded efforts" might step into the breach and undertake genomic sequencing projects of their own, "leading to incomplete data sets of marginal quality to which the scientific community has limited access."[11] With sequencing costs declining, a private company could "just set up business" and produce a rough sequence, with the goal of "skimming" it for genes and other useful things, for perhaps about $100 million—comparable to the investment that SmithKline Beecham had made in HGS.[12]

Meanwhile, how quickly to effect the transition remained a matter of debate, not least because everyone expected that massively scaling up sequencing would require winnowing down the number of genome centers—perhaps to a literal handful. Ollie, for example, argued that the fact that HGP money was being spent "in lots of places doing lots of things" had created an interest group that will "not benefit from the transition." Funds will have to be concentrated very heavily, he argued, not because the technology demands centralization but owing to a shortage of human resources, especially at the center director level. The United States would be "hard pressed to field five really strong sequencing centers."[13]

Choosing a Sequencing Strategy

As the HGP began to shift its focus to sequencing, questions about the sequencing strategy for completing the human genome took on new urgency. The projects to sequence the genomes of yeast and the worm had used a "clone-by-clone" or "hierarchical shotgun" sequencing strategy that involved a mapping step and a sequencing step. The mapping step entailed creating a library of partially overlapping clones from the target genome and then joining them into contigs, forming a contig map (as in figure 2.7). At the outset of the HGP, the clones used in contig maps were cosmids or YACs, which bacterial artificial chromosomes (BACs) replaced in the mid-1990s. The map provides a set of clones that, taken together, completely "cover" the genome yet overlap minimally. Each BAC clone is sequenced separately by breaking it randomly into small fragments that are fed into the sequencing machines. Computers assemble the reads from each clone into a continuous sequence of that clone. Finally, the sequences of the clones are anchored to the genome and joined, building up a continuous sequence in a clone-by-clone manner (see figure 7.1).[14]

In the mid-1990s, an alternative to the clone-by-clone approach emerged—namely, the "whole-genome shotgun" (WGS) strategy. This approach used the same fundamental technology as the clone-by-clone

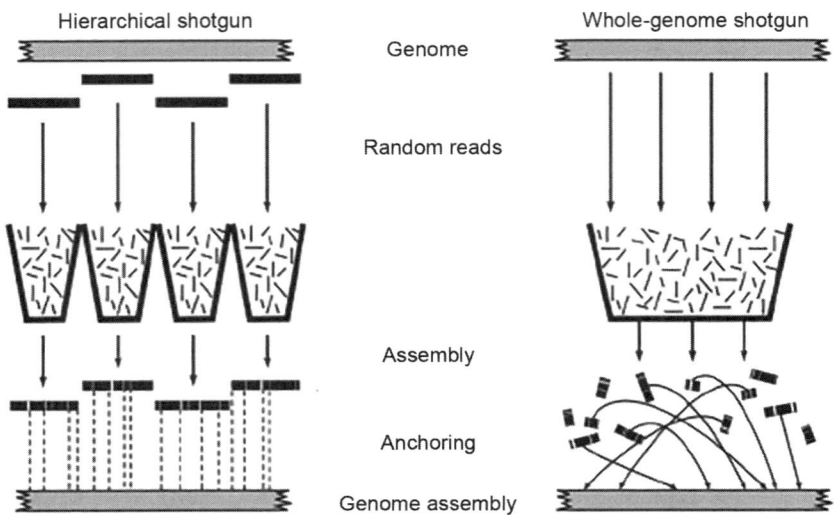

Figure 7.1
Clone-by-clone and whole-genome shotgun (WGS) strategies. The clone-by-clone strategy **(left)**, also known as the hierarchical shotgun strategy, begins with the construction of a set of overlapping BAC clones. Each of these clones is separately "shotgun sequenced"—that is, shattered into small pieces, sequenced, and assembled into contiguous sequence. The WGS strategy **(right)** begins by shattering the entire genome into small pieces, sequencing them, and assembling all of them at once into contiguous sequence. This strategy skips the steps needed to map the BACs, but the assembly process is more complicated. The assembly step in the clone-by-clone approach benefits from compartmentalized assembly. A typical BAC contains about 200,000 base pairs—or about $\frac{1}{15,000}$ of the human genome. Assembling each BAC thus entails working with many fewer fragments than must be assembled in WGS strategy. Each BAC also has a known location on the genome, providing additional information that the clone-by-clone strategy can use to assemble and anchor contigs to the genome. *Source:* Waterston, Lander, and Sulston 2002.

approach—sequencing small fragments of DNA with automated sequencing machines. However, the WGS strategy used the machines in a different way, skipping the mapping step that is key to the clone-by-clone approach. The WGS strategy entailed breaking the *entire* genome randomly into a single batch of small fragments, which were then sequenced, producing a vast number of short reads that cover the genome. Computers assembled the reads into contiguous sequence by identifying overlaps. Skipping the mapping step saved work but at the cost of increasing the complexity of the assembly process (see figure 7.1).

At TIGR, Craig Venter had used the WGS strategy to sequence *H. influenzae*, a bacterium, which became the first freestanding organism to have its genome completely sequenced.[15] The TIGR group shattered the entire 1.8 million base pairs of the *H. influenzae* genome into small fragments and sequenced them. Next, computers compared the reads to look for overlaps and joined overlapping sequences to reconstruct the original sequence. The method had worked well for this small genome, and Venter soon began sequencing additional microbial genomes using the WGS method. Whether this method could be scaled up to the 3-billion-base-pair human genome—some 1,600 times the size of *H. influenzae*—became a matter of debate.[16] One of the key issues concerned "repeats"—that is, DNA sequences that occur many times in the genome. Repeats are rare in bacterial genomes, but much of the human genome consists of repeats. When assembling a collection of short strips of sequence data into a continuous sequence, repeated sequences that exceed the length of the reads result in ambiguities about the correct order of the strips of sequence (see figure 7.2). Because the WGS strategy entails comparing a huge number of reads from all parts of the genome, repeats (which are troublesome enough in the clone-by-clone approach) pose especially serious problems for this technique. As a result, the WGS strategy might produce a final product containing many strips of sequence in the wrong order or a final product consisting of many disconnected strips of sequence—in other words, many errors and gaps.

Following Venter's success with *H. influenzae*, the HGP considered and rejected the possibility of switching to a WGS approach. In discussions held in 1996 and a published exchange in *Genome Research* in 1997, James Weber and Eugene Myers made the case for using the WGS strategy to sequence the human genome. Using computer simulations, they assessed various ways to perform a WGS of the human genome and concluded that the method would be cost effective and quicker than clone-by-clone sequencing.[17] Philip Green countered with an analysis that questioned some of the assumptions underlying the Weber–Myers simulation and concluded that the WGS strategy would produce an inferior-quality sequence with many errors and gaps that would be difficult to close. Green argued that

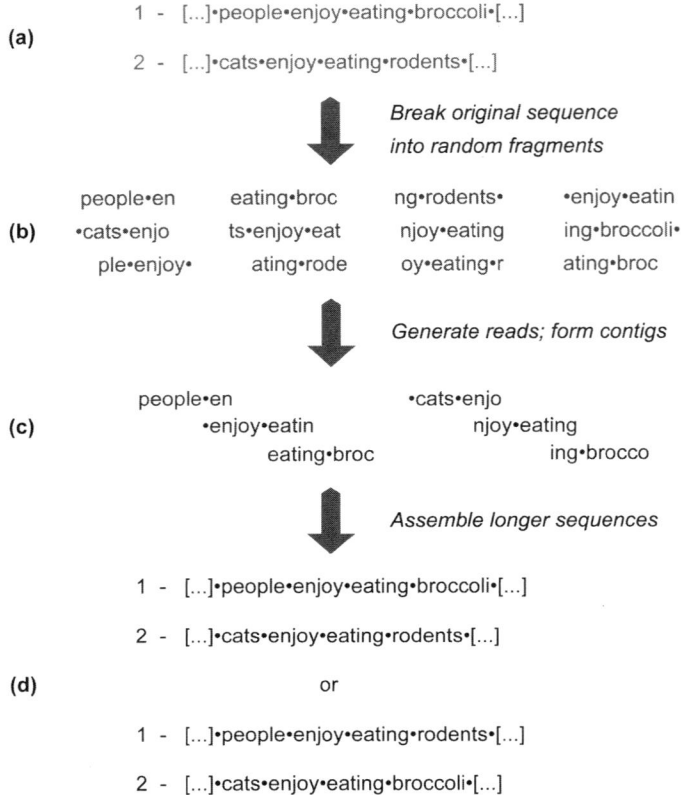

Figure 7.2
Repeated sequences produce ambiguities. DNA sequences that occur multiple times in a genome can produce ambiguities when sequencing. The basic principle is illustrated in greatly simplified form using a sequence of English-language characters. **(a)** Two strips of text from different parts of the same "genome" appear in 1 and 2. They are shown in gray to indicate that the sequences are initially unknown. Both of these sequences contain a "repeat": the 14-character text •*enjoy*•*eating*•. In this example, the maximum read length is 12 characters, which is shorter than the repeat. As a result, when these sequences are shattered into small fragments **(b)** and read and assembled into contigs **(c)**, none of the reads extends all of the way through the repeat. For this reason, the assembly process will yield multiple configurations **(d)**, only one of which is correct. Much more complicated ambiguities result from repeats that recur many times in a genome. Producing highly accurate, finished DNA sequences thus may require undertaking additional experiments and analyses to resolve ambiguities and obtain accurate assemblies. Illustration by Ranjit Singh.

cost-effectiveness had not been demonstrated, that the WGS strategy was risky, and that the case for changing course had not been made.[18]

The HGP leadership, which was committed to producing an accurate and complete sequence, found these arguments persuasive and rejected the WGS strategy. However, some genome project leaders continued to worry that a company might try to launch a "quick and dirty" WGS project designed to rapidly produce a low-quality sequence, which would allow the company to "skim" the genome for interesting things with the goal of capturing control over a great deal of valuable information.

Announcing the New Company, May 10–13, 1998

The founding of Celera in 1998 occurred in this context, and a struggle to shape the news coverage unfolded around the announcement of its existence.[19] According to Venter's autobiography, published nine years later, the discussions leading to the creation of Celera began in late 1997, when the head of ABI at Perkin-Elmer, Michael Hunkapiller, began to explore with Venter the prospects for sequencing the human genome using a new automated sequencing machine that the company was developing. Perkin-Elmer was willing to invest $300 million in such an effort, and the parties created a new company with Venter at the helm.[20] Because the company did not yet have a name (the name "Celera" was not announced until July 1998), people initially availed themselves of such descriptors as "the New Company" or "the Venter/Hunkapiller proposal" when discussing it.

The Announcement

The announcement of the formation of the company was carefully staged, using the selective and carefully timed revelation and concealment of information central to public-relations strategy. The company initially preserved a monopoly on information about its existence and plans, keeping the news under wraps until May 10, 1998, two days before the annual Cold Spring Harbor meeting.[21] The New Company quietly invited a handful of well-known scientists to join its advisory board. Shortly before the announcement, a few HGP leaders (including James Watson, Sanger Centre director John Sulston, NHGRI director Francis Collins, and the Wellcome Trust program director Michael Morgan) learned that Venter was forming a sequencing company. On Friday, May 8, two days before the dramatic announcement of the company's plans, Venter and Hunkapiller personally met with Collins and Harold Varmus, the director of the NIH; it was then that NIH officials first learned that Perkin-Elmer, the leading manufacturer of sequencing machines, was involved.[22]

Venter describes the company's strategy of offering "the story on a plate to a reporter who we thought could get the space to do it justice"—namely, Nicholas Wade of the *New York Times*. The plan was for the New Company to "issue a press release before the markets opened on Monday, May 11, 1998," but Wade's story would appear in the Sunday *New York Times* on May 10.[23] Wade's exclusive article—headlined "Scientist's Plan: Map All DNA within 3 Years"—appeared on the front page, and as the first article to introduce the new company to the world, it played a key role in framing the meaning of the company's existence. Wade articulated what I call the "company-takeover" narrative: namely, a story of an upstart private company, led by a pioneering scientist, that sought to upstage the government's genome program by finishing the job in half the time for one-tenth the investment. The article began: "A pioneer in genetic sequencing and a private company are joining forces with the aim of deciphering the entire DNA, or genome, of humans within three years, far faster and cheaper than the Federal Government is planning. If successful, the venture would outstrip and to some extent make redundant the Government's $3 billion program to sequence the human genome by 2005."[24]

Wade's article presented a number of entities in ways that supported this plotline:

• *Dr. J. Craig Venter*: "president of the nonprofit Institute for Genomic Sciences" and "a pioneer in genetic sequencing."
• *The company's ambitious style*: "The proposal to substantially complete the human genome in three years would seem extreme hubris coming from almost anyone but Dr. Venter. But other experts deemed his approach technically feasible."
• *The company's method*: "a new sequencing strategy" that Venter had developed. The machines would provide "the means to execute" Venter's "new sequencing strategy."
• *The new sequencing machines*: "a new generation of sequencing machines … so fast that the whole human genome could be completed far sooner and 10 times more cheaply than envisaged by the National Institutes of Health."

- *The company's expected scientific output*: "the" human genome sequence—
"the three billion letters of human DNA."
- *The speed and cost of the plan*: The company would sequence the genome
"within three years, at a cost of $150 million to $200 million."
- *The company's date for completing the human sequence*: 2001.
- *The government's genome program*: a "$3 billion Federal program" presently
"at the halfway point of its 15-year course, and only 3 percent of the
genome has been sequenced." "Although the program has had many successes in pioneering a daunting task, serious doubts have emerged as to
whether the universities can meet the target date of 2005."
- *The effect of the company on research*: "The project could have wide ramifications for industry, academia and the public because it would make
possible almost overnight many developments that had been expected to
unfold over the next decade."
- *The company's disruptive effects*: "Congress, for instance, might ask why it
should continue to finance the human genome project ... if the new company is going to finish first." The company would create a database that
"seems likely to rival or supersede" the government-funded GenBank.
- *The government program's response*: The company would lead the government's genome project to shift from human sequencing to sequencing the
genomes of mice and other organisms. Varmus and Collins, Wade reported,
"expressed confidence that they could persuade Congress to accept the
need for this change in focus."[25]

This company-takeover narrative—with Venter cast as a bold pioneer,
the government program as sluggish and expensive, the company's plan as
highly credible, and the company's creation as completely transforming
genome research—became one of the main ways that journalists, along
with Venter, Celera, and others, initially framed the unfolding events.

Although Wade's article centered on the New Company's challenge to
the government program, it also contained several conciliatory statements
from company principals, suggesting their intention to build a cooperative
relationship with the HGP. The article reported that Venter wanted to "work
closely" with the NIH and quoted him as saying that he did not want to
evince "an in-your-face kind of attitude." The article also suggested that
"public concern" might develop about the prospect of one company controlling so much genomic information, although it did express some doubt
about whether the New Company could use patents to "gain a significant

lock" on the genome. It quoted bioethicist Arthur Caplan, who asked whether the moral and legal questions raised by sequencing the genome could be addressed "if the largest scientific revolution of the next century is going to be done under private auspices."[26]

The New Company's press release, which appeared the next day, announced the formation of the company and claimed that it would "substantially complete the sequencing of the human genome in three years": "Using breakthrough DNA analysis technology being developed by Perkin-Elmer's Applied Biosystems Division, applied to sequencing strategies pioneered by Dr. Venter and others at TIGR, the company will operate a genomics sequencing facility with an expected capacity greater than that of the current combined world output." The company, the press release continued, planned to extract biological knowledge with "significant commercial value" from genomic data, discovering new genes, finding polymorphisms to use as markers in research and diagnostics, and creating databases for use by the scientific community. The press release also emphasized speed—"the new company can provide this information more rapidly and more accurately than is currently possible"—and said that the HGP, which had begun in 1990, had so far developed maps and sequenced only 3 percent of the human genome.[27]

As is typical among high-technology startup companies, which some analysts have called a world of "story stocks" that acquire much of their value by presenting compelling narratives, the press release was aimed not only at the media but also at investors. It directed Perkin-Elmer investors to one telephone number and referred journalists to the public-relations firms Noonan/Russo (serving Perkin-Elmer) and Hill & Knowlton (serving Venter and TIGR).[28] A second press release by Perkin-Elmer announced its new sequencing machine.[29] As is typical of the genre, neither press release contained much technical information. They also said nothing about the sequencing strategies Venter had pioneered or about the company's business model. Importantly, however, the company did say that it would "make sequencing data publicly available to ensure that as many researchers as possible are examining it"—a formulation that made its plans seem, at least potentially, to be consistent with the HGP goal of making sequence data public. The terms of access—and other important aspects of the knowledge-control regime that the company envisioned—were not discussed.[30]

The US Genome Program's Initial Response

Responding to the New Company's sudden appearance on the public stage posed challenges for HGP leaders. Not only did the New Company get to launch the action at a time of its choosing and initially cast events in the company-takeover narrative, but the HGP leadership also had incomplete information about the company's technology and plans, making it impossible to produce an authoritative assessment of the situation. Moreover, the company-takeover narrative—which defined the HGP as an inefficient government program, not an exciting scientific quest—put the leadership of the US genome program in a rather delicate position, given the enthusiasm for unfettered markets and minimal government prevailing in the United States. The HGP leadership did not want to allow the New Company to take over human sequencing, not least because of doubts about the terms for making the data "publicly available"—especially in light of the prior experience with ESTs in which Venter had figured so prominently (see chapter 5). HGP scientists also expressed concern that the details of the company's plan remained sketchy, that its technical feasibility remained undemonstrated, and that no one outside of Perkin-Elmer had worked with the new sequencing machine. Yet simply dismissing the plan, which was spearheaded by a well-known genome scientist and backed by the financial and technical resources of the world's leading manufacturer of sequencing machines, might make the HGP appear rigid and defensive. For the NHGRI and the DOE, the US Congress, which at the time was controlled by Republicans with a strong antigovernment ideology, was an especially salient audience. In this context, the US agencies simply chose to "welcome" the New Company's plan as a positive development while politely raising questions about whether the proposed method would work and, if successful, whether the company's product would meet the HGP's goals for data quality and access.

On Monday, May 11, a press conference was held at the NIH, featuring Varmus, Collins, Ari Patrinos of the DOE, Venter, and Hunkapiller. John Sulston later called this joint press conference "the first of what was to become a series of bizarrely staged shows of unity."[31] Both the New Company and the government officials presented themselves as willing to work together. An account from one of my interlocutors who attended the event indicated that the message the NIH tried to convey was that the New Company's initiative was something that "should be embraced" if it played out as portrayed. *If* was the operative word: the intended message was that the new initiative *could* be a helpful supplement to the HGP *if* the sequencing strategy were actually to produce high-quality data and *if* the data were in fact made truly publicly available. Given these *if*s, the public program should take no hasty steps.[32] The core of this "wait-and-see" narrative was the idea that the company's method and data-access plans remained unproven.

Media coverage of the press conference presented a somewhat confusing mixture of the company-takeover and the wait-and-see narratives, along with a narrative about "commercial capture" of the genome (table 7.1). A *Washington Post* article titled "Private Firm Aims to Beat Government to Gene Map" began with the company-takeover narrative and made claims that cast private industry as much more efficient than government projects. It quoted William Haseltine, the head of HGS and veteran of the struggle over ESTs, who called the government's program a "gravy train" for the HGP's sequencing laboratories and faulted HGP leaders for their "failure" to enlist private industry. But the article also presented the wait-and-see narrative, suggesting that biotechnologists were divided on the question of whether the company would achieve its ambitious goals and quoting Collins as saying that "it would be vastly premature to go out and ... change the plan of our genome centers." The HGP would "remain on its current course for the next 12 to 18 months, by which time it will be clearer whether the project should change its approach." The *Washington Post* story also said that the development "raised fresh concerns about the prospect of the human genetic code being expropriated by entrepreneurs who plan to patent and sell access to the most medically valuable parts."[33] In contrast, a second article by Wade in the *New York Times* presented only the takeover narrative and used the past tense to suggest that the "takeover of the human genome project" was a fait accompli. Wade quoted Haseltine, who pointed to "serious problems of organization and management both at the Department of Energy and at N.I.H." The article also reported that NIH officials were preparing to "switch the focus from getting the sequence to the enormous tasking of interpreting it," perhaps by sequencing "the surprisingly similar genome of the mouse."[34]

Even HGP scientists found the news coverage of the press conference confusing. One scientist, JOHN, described how he missed the news on the Sunday and Monday when the story broke. When he finally did read the

New York Times accounts en route to Cold Spring Harbor on Tuesday, May 12, he found them baffling.

I guessed they [the New Company] meant whole-[genome] shotgun. I guessed that they meant looking for lots of polymorphisms quickly. I guessed that they weren't serious about closing the genome [i.e., completing it without gaps]. What shocked me was that NHGRI, from reading the newspaper, I got the impression that the leaders had said we're going to move to mouse and we're going to move to other things. And I said, "How can you retreat? There's so much information that you don't know here that we shouldn't be changing our plan." I met DIRECTOR OF EASTERN GENOME CENTER Tuesday night when I arrived here the day before for a premeeting. I confirmed the shotgun and learned NHGRI hadn't said these things about retreating.[35]

On Tuesday, May 12, the day before the annual international meeting of genome scientists was to begin, the heads of the largest American genome laboratories gathered in Cold Spring Harbor. Venter and Hunkapiller attended part of this premeeting gathering and presented their plan to the genome center directors. By all accounts, the New Company's proposal was poorly received. The center directors remained unconvinced that the WGS strategy would produce a high-quality sequence, and they doubted that the new company's terms of "public" access would be acceptable. Venter and Hunkapiller left Cold Spring Harbor before the annual meeting began.[36]

The Wellcome Trust's Response

News of the New Company's plan provoked intense discussions among HGP scientists about how to respond. Some of the most consequential discussions involved the Wellcome Trust, funder of what was then the world's largest genome laboratory, the Sanger Centre in the United Kingdom. In his account of these events published in 2002, Sulston describes how he and his colleagues worried that the US program was in jeopardy. What was needed, he believed, was a powerful statement that could counter the mounting perception that the public HGP was clumsy, inefficient, and no longer necessary: "Trying to get reporters to print the admittedly more complex analyses that we felt were being ignored was going to be an uphill struggle. We were learning fast that we would have to play the public relations game if we were to survive. But that didn't mean indulging in empty hype. What we needed was a big vote of confidence in the public project as a counter to Craig's [Venter's] hints that it was an expensive white elephant."[37]

Table 7.1

Three Narratives in the Initial News Coverage of the Human Genome Sequencing
Competition

Takeover	• "The sequencing of the human genome, a historic goal in biomedical research, was snatched away last Friday from its Government sponsor, the National Institutes of Health, by a private venture that says it can get the job done faster." (*NYT*) • "Now Government officials are scrambling to adjust to the stunning turn of events, saying that the task of interpreting the genome may begin much sooner now, and that there is every reason for Congress to continue to fund the project." (*NYT*) • "'There have been serious problems of organization and management both at the Department of Energy and at N.I.H.' together with internal dissension among the senior scientists involved, said Dr. William A. Haseltine." (*NYT*) • "It may not be immediately clear to members of Congress that, having forfeited the grand prize of human genome sequence, they should now be equally happy with the glory of paying for similar research on mice." (*NYT*) • "Scientists yesterday said they would form a new company in Rockville that aims to unravel the entire human genetic code by the year 2001, four years sooner than the federal government expects to complete a similar project." (*WP*) • "The privately funded enterprise, which backers said could be completed for perhaps one-tenth the cost of the government program, raised immediate questions about the relevance and future of the $ 3 billion, 15-year federal effort." (*WP*) • "William Haseltine yesterday called the government's program a 'gravy train' and faulted its leaders for what he described as a failure to enlist private industry." (*WP*) • "'This has to feel like a bomb dropped on the head of the Human Genome Project,' Haseltine said by telephone from Frankfurt. 'All of a sudden somebody is going to pull a $3 billion rug out from under you? They must be deeply shocked.'" (*WP*)
Wait-and-see	• Genome experts "seem to accept with little reservation that the abductors have a reasonable chance of making good on their claim to substantially complete the human genome, starting from scratch, in three years." (*NYT*) • "Some biotechnology experts not involved in the new company raved about the venture, saying it promises to generate enormous amounts of genetic data." (*WP*) • The company "will use a controversial approach called 'shotgun whole genome sequencing.' Instead of focusing on large pieces of DNA, this process decodes tiny pieces that later must be assembled like interlocking pieces of a jigsaw puzzle. Because of the added difficulty of dealing with so many small pieces, the resulting picture of the human genome is likely to be peppered with more and larger holes than that produced by the federal program, Collins said." (*WP*)

Table 7.1 (continued)

	• "The government considered switching to the approach that Venter will use a few years ago, Collins said, and 'roundly rejected' it as too problematic. *But Venter and others said recent technical improvements make the approach superior."* (WP)
Commercial capture	• *"Dr. Venter plans to enter his findings in a public database."* (NYT) • Biotechnology "companies have been granted scores of patents on their genetic discoveries, raising fears among some critics that a handful of companies will control the commercialization of a vast and potentially lucrative biological resource. Those fears arose again yesterday with Venter's announcement of his new project." (WP) • "'Even though they are promising public access, they control the terms and there is a history of terms being more onerous than is acceptable to most scientists,' said Maynard Olson." (WP) • *"Venter said that with the exception of perhaps 100 to 300 genetic sequences that he expects will show special commercial promise, the company will make all the genetic information available free to the world's scientists. 'It would be morally wrong to hold the data hostage and keep it secret,' he said."* (WP)

Examples are from the news coverage of the May 11, 1998, press conference in two leading US daily newspapers, the *New York Times* (*NYT*, Wade 1998a) and the *Washington Post* (*WP*, Gillis and Weiss 1998). Text displayed in italics contradicts the narrative.

On Wednesday, May 13, Sulston, along with Michael Morgan of the Wellcome Trust, made a presentation to a previously scheduled meeting of the Trust's board of governors that persuaded them to make such a statement in the form of a serious financial commitment.[38] That same day the Trust issued a press release saying that it would increase its funding of human genome sequencing by "£110 million over seven years, bringing the total Trust investment in the Human Genome Project to £205 million." The press release also said that "the Trust is concerned that commercial entities might file opportunistic patents on DNA sequence. The Trust is conducting an urgent review of the credibility and scope of patents based solely on DNA sequence. It is prepared to challenge such patents."[39]

Cold Spring Harbor, May 13–17, 1998

At the Cold Spring Harbor meeting, the assembled scientists extensively discussed the New Company, and many of them were concerned with the impression conveyed in the news coverage that the HGP was failing. Prior to the meeting, many genome scientists had acquired information about the New Company's plans mainly through media coverage, and my interlocutors also expressed frustration with the sketchiness of the details about the technical issues, about the company's business model, and even about the NHGRI's response. They attempted to make sense of the rapidly moving developments, using the sparse information in the news accounts to make educated guesses.[40] They speculated about the knowledge-control regime that would underlie the company's business model. How did the company plan to make money? Would it patent genes, as the positional cloning companies had done, such as Myriad Genetics, which had patents on two genes implicated in breast cancer?[41] Would it sell access to sequence data? Build databases of polymorphisms for use as genetic markers in research and diagnostics? Patent the valuable genetic markers known as "single nucleotide polymorphisms" (SNPs, pronounced "snip")? Sell access to a SNP database? An Associated Press story reported on May 11 that Venter had said that the company "would release its results every three months and would not patent any genes,"[42] but the genome scientists with whom I spoke expressed distrust of the New Company's promises to provide public access to its data, calling these assertions vague, worrying about delay or burdensome terms, and pointing out that company policies can change.

Some HGP scientists also speculated that the company probably planned to file patents on SNPs, which they regarded as "research tools" that should not be patented.[43] They also suspected that Venter's "new" method was "essentially the same" as the WGS approach that the HGP had rejected and predicted that it would produce a sequence with a huge number of gaps.[44] Some suggested that the company seemed to be deliberately obscuring the difference between what the HGP planned to deliver and what a WGS strategy would produce. They objected to "science by press conference" and called for a proper "scientific evaluation" of the company's plan. Some of my interlocutors also feared that the New Company might be trying to push the NIH out of the human genome sequencing race, perhaps by going to Congress to argue that the HGP was duplicating work being done in the private sector.

My interlocutors were also deeply concerned that influential observers, especially members of Congress, would misperceive the situation given their exposure solely to a mass-mediated version of events. As one genome scientist, KEVIN, put it in a May 15 interview,

I think even my mother would look at this and say, "Oh, wait, somebody's saying they're going to do the genome." I'm not sure—my mother may not be savvy enough to know the difference, and the congressmen may not be either, [between] what's an intermediate product versus the final product. "I think it sounds like a good short-cut." "We don't need to be spending all this extra money." And so you just have to be concerned about that, and we need to keep enthusiasm constantly up for continued good, healthy funding of the genome project. And if somebody can run out and get a shortcut, it may not dawn on them until several years later that, "oh my God, that wasn't really quite what we wanted."[45]

Part of the problem, according to my interlocutors, was that misleading media coverage conveyed the impression that what people had begun to call the "public genome project" was failing. Kevin presented the matter quite directly:

Kevin: This is a heavy amount of propaganda coming and publicity coming from them. And on top of it, it's a very harsh treatment by the press, and particularly the *New York Times*, of this whole thing. So it's not only that you have this scientific agenda that could be perceived as undermining the public [project], but … instead of having time to scientifically evaluate and discuss this, immediately there's been some major darts thrown at the organization in the execution of the public genome project. … "How come we're half way through the genome project and we only have 3 percent of the genome sequence?" I mean, this is a really stupid argument. … And it is just absolutely ridiculous. It was a planned program from the beginning. … And it's measured in milestones, and we … didn't even think we'd be this far along in sequencing now. And forget the fact that we've built physical maps. Forget the fact that we've built genetic maps. Forget the fact we've built infrastructure for sequencing and just focus on the 3 percent. This is just really naive, and it's just a stupid way to look at it. So it's a lot of publicity, and it's that type of publicity. …
SH: Have you seen, coming out of the New Company or the group, any kind of documents? Any kinds of papers?
Kevin: [disdainful] Press releases. Only thing I've seen.[46]

The organizers of the Cold Spring Harbor meeting spliced a special session on the Venter/Hunkapiller announcement into the meeting agenda. No official representative of the New Company was present, so the organizers asked a well-known HGP scientist, MATTHEWS, to summarize what the HGP leadership knew about the proposal based on news reports and the May 12 meeting that Venter and Hunkapiller had held with genome center directors. This session offered the leadership an opportunity to build morale, challenge the idea that the project had been snatched away, and build a collective response based in shared framings of the situation. But there were also issues that they sought to leave backstage. Even before the Cold Spring Harbor meeting had officially started on May 13, genome project leaders had begun vigorous closed-door discussions about how to respond—both in public statements and in their scientific strategy. An acceptable response would have to maintain the HGP's commitment to its original goals and address the possibility of a campaign to cut funding of a "wasteful" public program. But the leadership was far from making final decisions about how to proceed. Moreover, anything "officially said" at Cold Spring Harbor had a decent chance of appearing in the press. The registrants for the meeting included several science journalists as well as a few employees of Perkin-Elmer and TIGR. The scientists leading the HGP—and especially the representatives of funding agencies such as NIH—therefore could not avoid performing in an official role. Unlike a press conference, this session did not address journalists directly, but the visibility of events made the session into something different from an ordinary discussion at a scientific meeting.

Grace Auditorium was packed, and a few people were sitting on the floor. Matthews prefaced his remarks by framing his role in the session as "unusual," noting that his charge placed him in a "somewhat awkward" situation that offered "some temptations," but he nonetheless would try to be "dispassionate" and present the proposal fairly. With the audience thus notified that he had adopted a self-consciously neutral persona, distinct from his everyday self, he proceeded in a just-the-facts mode, describing some technical details about the WGS approach and Perkin-Elmer's new sequencing instrument. Some of the information in Matthews's description had previously appeared in news accounts and press releases announcing the New Company, but other information had not. Perkin-Elmer's press release on the new instrument, for example, merely referred to it as "ultrahigh throughput," whereas Matthews had sufficient quantitative details to shed light on its theoretical throughput (96 capillaries/run × 500 base pairs/capillary × 3 runs/day = 144,000 base pairs/day)—very fast by standards in 1998. For the experts in the room, such details were informative but raised additional questions. How closely would the instrument—which was still under development—approximate its theoretical throughput? What was its error rate? How much technician time was required to operate it? Matthews's presentation also addressed the data-release policy, and he noted that the company currently planned to release data quarterly and that the company said it expected to patent 100 to 300 genes. Commenting on the $300 million investment, Matthews noted that "given that amount of money, there has to be a business plan"; then, apparently concluding that he had stepped outside his neutral role, he added, "Sorry, that was an editorial comment."

Matthews was immediately followed by a representative of the Wellcome Trust, who said that before news of the New Company broke, the Trust was already discussing the idea of increasing its financial commitment to the HGP. In response to the founding of the company, the Trust decided to publicly announce a new, scaled-up commitment. The resulting press release was issued with the "enthusiastic support" of the "entire" Wellcome Trust Board of Governors. He then proposed to read the press release "into the record of this meeting." When the Trust's representative had finished, he received extremely enthusiastic applause.

An open discussion followed these presentations. Genome project leaders from both the United Kingdom and the United States reiterated what had become the official line of "welcoming" the new initiative while also stressing the need to complete the entire sequence and not be derailed from that goal. An American HGP leader stressed that the company's sequence data would be useful but emphasized that the noble goal of decoding the "Book of Life" could not be satisfied with an incomplete view. In an allusion to the expectation that the New Company's WGS sequence would have many gaps, he said that: "We cannot be satisfied by disconnected paragraphs in a bag without order and connection." He also said that the press accounts reporting either that the HGP would pay no attention to this development or that it would move on to the mouse were not true. "We should be energized by this announcement," he said. "An additional $300 million! Let's see if we can use it to speed up the project."

One questioner sought more information about the Wellcome Trust's plans to challenge the patentability of DNA. The representative of the Trust replied that it would make a serious effort to understand the law and to consider challenging the patents, adding that the Trust has the resources needed. Some questions stayed close to technical issues. For example, one member of the audience asked whether it might be possible for HGP scientists to use simulation to understand what this proposed project would be able to do. He was answered from the floor by Ted, a mathematician, who argued that such a simulation would be hard to do but that the WGS strategy was likely to produce an incomplete sequence with many gaps.

When the discussion turned to the details of data release and patenting, audience members began calling out questions: "Would the company patent SNPs?" "Mike Hunkapiller and Craig Venter said they would not, right?" "Can we rely on their quarterly data-release policy?" "They said they would not release the unprocessed 'trace' data produced by the machines, right?" Orderly turn taking began to break down. GREG, a scientist with connections to the New Company, jumped up from the audience, and he was given the floor with the words "Let's hear from someone who knows what's going on." Greg said that many details remain to be decided: The people who are launching the New Company do not have a really concrete plan of exactly what to do. It is a company in process, not in concrete. Part of the business plan includes early release for a price with later release to all for free. Be really careful making judgments, he stressed, because they are still working out the details themselves. When he was done speaking, the session had reached the end of its scheduled time and was adjourned.

Competing Laboratories, Competing Narratives

The Cold Spring Harbor meeting ended two days later, with the stage set for ongoing competition between the HGP and the New Company, which was named Celera two months later, in July 1998. Following the Wellcome Trust's commitment of extra funding, a "competition" narrative, focusing on a struggle between two groups—one "public" and one "private"—took shape. Some of the media accounts used war metaphors. The *Montreal Gazette* reported that Venter had "secured $300 million of private funding to launch his assault," saying his "chief weapon" was speed and describing the Wellcome Trust funding as "retaliation."[47] Wade published a third *New York Times* article on May 17 headlined "International Gene Project Gets Lift." After saying that the $3 billion project had been "upstaged" by a private company, the article announced that the Wellcome Trust had "stepped into the fray in an effort to maintain the impetus of the publicly financed program and to prevent the human genome sequence from falling under the control of a private company." It quoted Michael Morgan as saying that "to leave this to a private company, which has to make money, seems to me completely and utterly stupid." Morgan also said that the Trust, with assets of $19 billion, could finance the sequencing of the entire human genome: "if we had to and if we wanted to, we could do it."[48]

As the competition between the HGP and Celera played out over the next few years, actors from both camps occasionally publicly declared their desire for cooperation, but many observers interpreted this sentiment as a thin veil over irreconcilable goals. At a congressional hearing on the HGP in June 1998, for example, Francis Collins, Ari Patrinos, and Craig Venter all expressed a desire to collaborate. In Patrinos's account, making the two efforts complementary would depend on the company providing "prompt and complete" access to its raw data. In exchange, the HGP would produce a fully mapped "draft" or "scaffold" version of the genome, on top of which the company's sequence could be assembled. In Patrinos's vision, the HGP would later fill in the sequence in greater detail.[49] Negotiations about the possible terms of collaboration took place at several points during the next few years. As one might expect, the two sides proved unable to agree on a plan along the lines of what Patrinos imagined, not least because the knowledge-control regimes that they sought to emplace could not be reconciled.

Throughout the spring and summer of 1998, HGP scientists continued to debate whether the project should significantly change its strategy to respond to the creation of Celera. The commitment to the original HGP goal of putting a high-quality sequence in the unrestricted databases remained firm, but some influential leaders, including Eric Lander of the Whitehead Institute, strongly argued that the HGP should speed up production and create an "interim" "draft" sequence along the way.[50] Better to divide the HGP's production goals into "draft" and "final" versions than to allow Celera to proceed unchallenged and produce a quick-and-dirty sequence, thus capturing control of valuable data and claiming priority for "completing" the first human sequence. When news of the New Company was announced, the US program had already been working on developing a new set of five-year goals to supersede the ones for the 1993–1998 period, and in September the HGP announced a significant change in course: the HGP would produce a "working draft" by the end of 2001 and "complete an accurate, high-quality sequence" by the end of 2003—two years ahead of the original schedule.[51] The US genome program expected "to contribute 60 to 70% of this sequence, with the remainder coming from the effort at the UK's Sanger Centre and other international Partners."[52] The HGP sharply distinguished the "working draft" (also referred to as an "intermediate product") from the "finished" sequence that it remained committed to produce.[53] The HGP leadership expected that the working draft, like Celera's WGS sequence, would have many gaps and problems and was not intended to be an acceptable final product, but the draft would be useful for finding genes and many other research purposes.

The change in strategy resulted from many interrelated considerations, such as shoring up congressional support for the US genome program; making a valuable "intermediate product" available to the scientific community; defining an interim milestone to prevent observers from equating Celera's low-quality sequence with the "finished" sequence that the HGP would eventually produce; and, not least, ensuring that the sequence ended up in such databases as GenBank, free of access restrictions and unencumbered by intellectual-property claims.

Over the next two years, the HGP and Celera engaged in a widely publicized competition. During the competition, my interlocutors included allies of both camps as well as a few people who expressed discomfort at what one called the mutual "bad mouthing" taking place, but I had no access to the internal process of decision making by the leadership of either camp. That information control was tight at the time was not surprising to me given the high stakes and levels of investment, and I did not attempt to gain access to the sequestered spaces in which real-time decisions took place. Participant accounts (including Craig Venter's autobiography, John Sulston's book, and the authorized, inside account of Celera by science writer James Shreeve) describe negotiations both within Celera and the HGP and between them.[54] All of these accounts discuss sensitive meetings, backstage maneuvering around possibilities for collaboration, internal disagreements among each side's principals, and unexpected developments.

As the events unfolded, strategies further changed. In March 1999, the NHGRI announced (to Celera's surprise) that the HGP would further accelerate its sequencing and thus complete a draft sequence in 2000, a year ahead of Celera. In response, the company decided that it would download the HGP's raw sequencing information—which the HGP's laboratories made immediately available under the Bermuda Principles—and incorporate it into its WGS strategy. Celera launched a stock offering in April 1999, and in the ensuing months news coverage affected the price of its shares, sometimes positively and sometimes negatively. As the sequencing progressed, both sides issued press releases announcing that various milestones had been attained.

Scaling Up

As the competition between HGP and Celera unfolded, both sides made large investments in scaling up sequencing capacity. Celera's vision of producing a WGS sequence of the human genome in three years was predicated on the belief that the new sequencing instrument introduced by ABI, a Perkin-Elmer division, would work well in practice. As things turned out, genome scientists found the machine, marketed as the ABI Prism 3700, very effective. The ABI 3700s dramatically changed the sequencing process. All previous sequencing machines—such as the ABI 377, which had dominated the market—performed electrophoresis using traditional "slab" gels enclosed between glass plates. With the 377s, the rate-limiting step in generating reads had been working with slab gels. The government-funded facility we toured in chapter 2, Sequencing Center, had arranged its workflow and physical space to facilitate the meticulous tasks of washing and drying plates, carefully pouring gels, and loading DNA samples into small indentations in the slabs. The 3700s eliminated slab gels by performing electrophoresis in thin plastic capillaries containing the gel, neatly enclosed and ready to use. No gels had to be poured, no plates had to be cleaned, and loading DNA samples was greatly simplified. After seeing how well the new machines worked, Sequencing Center chose to retire its perfectly workable 377s and convert entirely to the 3700s. HGP laboratories invested heavily in the new instrument, and so ABI, like an arms manufacturer supplying both sides in a war, was soon selling to Celera's principal competitor the $300,000 machines and the patented reagents needed to use them.

For the HGP, scaling up required concentrating resources. The NHGRI made hard choices and directed the bulk of its funds to three sequencing operations: the genome centers at Washington University (led by Robert Waterston), the Whitehead Institute (led by Eric Lander), and Baylor University (led by Bruce Roe). The DOE consolidated the sequencing operations at its national laboratories into a single California facility operated by the Joint Genome Institute. The Wellcome Trust expanded its support for the Sanger Centre, and these five sequencing operations became known as "the G5," which together constituted the core of the International Human Genome Sequencing Consortium. Meanwhile, Celera built the world's largest sequencing facility. An investment bank report on Celera in 1999, which classified the company as a "speculative strong buy," described how the company had assembled: (1) an *unprecedented gene sequencing factory* of 300 ABI Prism 3700 DNA analyzers that can sequence in one month more DNA than has been added to the GenBank database ever"; (2) "the world's largest private parallel processing supercomputer facility"; and (3) "vast human expertise in molecular biology and software engineering." The report predicted that Celera "will become the leading supplier of knowledge regarding genomics and molecular biology, a revolutionary market that should experience exponential growth."[55]

* * *

While the HGP and Celera competed to scale up and produce data, supporters of Celera and scientists associated with the HGP produced opposing narratives of events. The main narrative that each group advanced formed a relatively coherent structure, with the plot and nature of the characters and moral stakes fitting together in a coherent story (table 7.2). Both narratives fitted with well-established moral discourses about the virtues and vices of societal institutions. Celera's story pitting a fast-paced company against a sluggish government bureaucracy neatly invoked the familiar discourse (especially prominent in the United States) that depicts the private sector as efficient and the government as ineffectual. Celera incorporated an emphasis on speed into its public identity.[56] Its very name was "derived from the word 'celerity' which means swiftness of motion" to symbolize the speed with which the new company planned to sequence "the complete human genome."[57] In contrast, the narrative depicted government inefficiency as manifested not only in lax management and organizational inertia but also in the HGP's commitment to a "slow" sequencing method.

HGP supporters produced a narrative that framed the competition as a struggle between a group of scientists committed to familiar discourses of scientific virtue, on the one hand, and a company that threatened the core principles of science, on the other. One strand of the story cast the HGP as defending the scientific community's virtues of openness and commitment to truth against a company seeking to transform the

Table 7.2

Narrating the Competition

Celera Supporters	HGP Supporters
The competition is between a fast private company and a slow, expensive government program.	The competition is between open and disinterested science and a closed, profit-driven company.
Celera will produce a "complete" sequence before the government project will finish its sequence.	The sequence that Celera plans to produce will not meet the HGP definition of "complete."
The human sequence is of such great scientific and commercial value that it must be available as soon as possible.	The human sequence is of such great scientific value that it must be available to all. It is humanity's common heritage, the Book of Life, the language of God.
Speed matters most.	Quality matters more. Unrestricted access matters most.
To pursue excessive quality is unnecessary and wasteful.	To settle for a quick-and-dirty sequence is unacceptable.
Celera will make the sequence publicly available.	Celera's promise to make the sequence publicly available is vague. Any access restrictions at all are unacceptable.
The WGS strategy will produce a high-quality sequence.	The WGS strategy will not produce a high-quality sequence.
The government program's response to Celera is motivated by self-preservation.	The government program's response to Celera stems from its commitment to scientific values, especially openness.

genome—an object of unprecedented scientific importance, sometimes figured as quasi-sacred—into a commodity. HGP scientists contrasted the genome project's imaginary of a genomic future of open science, exemplified by the HGP's commitment to providing unrestricted access to the human sequence, with the likelihood that Celera would hold sequence information behind a corporate firewall, making it available only under restrictive terms. HGP scientists also faulted Celera for doing "science by press conference" and for making assertions about the effectiveness of its WGS strategy without supporting evidence. The narrative advanced by some HGP supporters also tapped into the familiar discourse about companies whose quest for profit conflicts with the larger public interest: Celera's financial goals would require it to restrict access to sequence data, which would delay the progress of science.

Each of the competing narratives advanced a set of claims that together formed a self-reinforcing package. Celera and its supporters presented the rapid delivery of data, even if partial, as the most important virtue in genome sequencing because of the opportunity cost of delay—an argument that (not coincidentally) echoed a discourse of speed familiar from the earlier struggle over the cDNA strategy and ESTs. Some Celera sympathizers argued that filling the last gaps in the human sequence would reach a point of diminishing returns. Being too meticulous was not worthwhile; the point was to produce valuable data, not to build a "cathedral." Peter, a scientist sympathetic to Celera, told me that once "you've got the bulk of the information that you need," problems more interesting than closing gaps would emerge. He expected that "if it's not tasked out to industry for a reasonable price," completing the sequence would just be left undone. "And it's OK. I don't need the cathedral. I'd rather have the next mouse. I'd rather have the next chimpanzee. I'd rather have it to the same sort of level of completeness. That's where the information's at."[58]

In contrast, scientists associated with the HGP stressed the importance of delivering a top-quality final product. They described the human sequence—along with the sequences of other organisms—as a resource that would be central to biological research for a long time. These scientists recognized that a WGS sequence would be of considerable value, and they were only too aware that even the most painstakingly constructed sequence would have some errors and would encounter unsequenceable regions, producing gaps. But they believed that the WGS strategy fell far short of both achievable and desirable levels of completeness and accuracy.

Even more important were different views about the control of sequence information. Although Celera's business model did not emerge all at once, it was easy to argue that the company's goal of establishing itself as the leading supplier of sequence information had to depend on capturing control and at least temporarily restricting access. Some analysts believed the company's unprecedented sequencing and informatics capacity would create significant competitive barriers, allowing it to profit by selling access to sequence databases, human SNPs, agricultural genomic data, and comparative genomics information.[59] By late 1999, several large pharmaceutical companies had subscribed to Celera's databases.[60] Celera supporters saw nothing wrong with charging fees to subscribe to databases. Why not make the production of sequence information into a commercial operation?

Many inputs to biological science are produced by the research-supply industry, so why not genome data? Industry should supply sequences, and academic science should move in other directions—for example, to functional genomics studies aimed at understanding the operation and interaction of genes. Instead of embracing the opportunity to have a company take on the chore of sequencing, as one Celera sympathizer put it, the large genome centers decided to oppose the company because it was corporate, to suggest it was evil, to fight it because they did not want to be closed down.[61]

To many HGP scientists, however, the human genome sequence was no ordinary input to the research process on a par with a commercial enzyme or an advanced instrument. It was special—although precisely why varied among HGP scientists. HGP scientists uniformly spoke of the enormous value of the human sequence to science and medicine, but some of them also invested it with special symbolic and moral significance. The genome was sometimes linked to the sociotechnical imaginaries of a global humanity—a common heritage that belongs to everyone. Some scientists also invoked the mythic or the sacred (the "Book of Life" or the "language of God").[62] The objection to Celera's vision was not that the NIH would have to *pay* for sequence information (the government was already, after all, spending a great deal to produce that information); the objection was to *constituting an agent*—more accountable to shareholders than to the scientific community—with the power to restrict access to the human genome and other fundamental genomic information. For Celera to capture control over sequence production, they argued, would expose scientists in both universities and in life science companies to many potential liabilities, ranging from high prices to delays in access to reach-through agreements akin to those that HGS had imposed during the EST dispute. Such worries led a group of large pharmaceutical companies, along with the Wellcome Trust, to found the SNP Consortium, a project to fund HGP laboratories to produce a public database of SNPs, much in the way that Merck had supported launching a public EST database. Just as the public EST database eroded the value of HGS's EST database for leveraging collaborations, freely available SNPs promised to diminish the value of Celera's SNPs.

Some HGP scientists objected to Celera's use of HGP data in part because that use created a catch-22 situation. Because the HGP laboratories were releasing their sequence data daily under the Bermuda Principles, Celera could immediately incorporate those data into its analysis, but the reverse was not the case. The competition, according to some HGP scientists, resembled a baseball game in which every time you score a run, you must give a run to the other side. Celera sympathizers pointed out in response that the information had been released with no restrictions, which made it perfectly legitimate for the company to use it.

The symmetry in narrative coherence did not mean that each side's account enjoyed similar degrees of uptake. Throughout the competition, HGP scientists perceived themselves as being at a disadvantage in transmitting their message via mass media. Celera's plotline seemed plausible and inspiring to observers who identified with US business and political elites, which often express commitment to the discourse of private-sector efficiency. The story of a nimble company and maverick scientist taking on a stodgy government bureaucracy also fit nicely with prevailing media culture. Journalists figured Venter as the personification of the spirit of Celera—a bold, brash entrepreneur and brilliant genome scientist, a fiercely competitive individualist, and a long-distance yacht racer. News articles treated him as something approximating the scientific equivalent of a controversial Hollywood celebrity.[63] For example, a *New York Times Magazine* story in August 1998 presented a front-page headshot of Venter and reported that he had recently "elbowed his way to the center of what may turn out to be the most significant race in the history of modern medicine." "It was as if private industry had announced it would land a man on the moon before NASA could get there. As if an upstart company intended to build the first atom bomb." No leader of the public project was treated as a similarly colorful figure. The same article identified Francis Collins by his bureaucratic title, "director," and contrasted their personalities: "Venter is ever restless and impatient; Collins is methodical and precise (although, it should be said, he rides a red Honda Nighthawk motorcycle to work every morning)."[64]

HGP scientists saw their message about defending scientific openness as relatively unfamiliar to nonscientists and less exciting to journalists than a David-and-Goliath assault on a government program. They complained that the quality differences between a WGS sequence and the high-quality sequence that they planned to produce were too complex to describe to lay audiences and contended that Celera's public statements had obscured the differences.[65] They also suggested that many observers lacked the background to grasp the vague and potentially impermanent nature of Celera's promise to provide "public access" to the human sequence.

The competition to shape the narrative continued, expressing the themes described here, until it was ceremonially—if superficially and temporarily—resolved for the celebration of the HGP's "completion" of the human genome sequence held by President Clinton and Prime Minister Blair on June 26, 2000.

Ceremony and Celebration, June 2000 to April 2003

At the Cold Spring Harbor meeting in May 2000, there was not only a palpable sense that the HGP was drawing to a close but also a widely shared expectation that the competition between the HGP and Celera would continue. The HGP scientists gathered there expected that two versions (albeit incomplete ones) of the human sequence would soon be "done." By the end of 1999, both camps had generated enormous quantities of sequence data. Celera had decided to test the WGS method on the *Drosophila* genome, which was about 10 percent the size of the human, and in late 1999 the company declared that the assembly had worked (although critics pointed to its numerous limitations).[66] The HGP announced in December 1999 that it had sequenced chromosome 22, thus producing the first sequence of a human chromosome. During the first months of 2000, both Celera and the HGP reached various milestones.[67] Celera announced that it had finished sequencing and had begun to assemble its WGS, using its own data in conjunction with the HGP data it had downloaded from GenBank. HGP scientists were confident that their own working draft—by definition an incomplete, intermediate product—would be declared finished very soon. At Cold Spring Harbor in May 2000, scientists sympathetic to both sides told me that they expected an intense and highly mathematical debate about the relative quality of the Celera and HGP data during the year ahead.

The notion of "completing" "the" sequence of "the" human genome is conceptually slippery, owing in part to variation among human genomes and in part to flexibility in constituting a practical notion of "finished." (How many gaps are there? What percentage of bases are miscalled? How much do gaps and errors matter?) In championing speed over completeness, Celera had set aside questions about the acceptable number of gaps in its sequence. In contrast, the HGP had constructed a set of quality goals that included the goal of long-range continuity with a minimal number of gaps and an error rate of less than one mistake in every 10,000 bases.[68] However, the concept of a "working draft" reopened the interpretive flexibility of "complete," leaving it unclear when the HGP's draft would be done enough to actually call it "done." But if the precise meaning of completeness was unavoidably fuzzy, a closely related issue was anything but ambiguous: the HGP needed to produce a durable, collectively acknowledged truth: the *fact* that it had completed the human genome. Indeed, bringing the genome project to a successful conclusion—with the sequence *finished*—was critically important to a variety of entities. Not only did the International Consortium, the HGP leadership, and the funding agencies need to establish that they had achieved their goals, but the Clinton administration also wanted to produce a dramatic public demonstration of the HGP's success.

Yaron Ezrahi has persuasively argued that technological forms of instrumental action serve to undergird the political legitimacy of modern democratic states, especially in the case of the United States. On the one hand, technological achievements not only provide direct instrumental value but also serve as a demonstration of the efficacy of state action and at another level as a metaphor for the efficient operation of the machinery of governance.[69] On the other hand, as Sheila Jasanoff has pointed out, technological failures may cast doubt on the state's ability to govern effectively, especially if governmental responses are deemed inadequate.[70] The competition between the HGP and Celera was often interpreted as bearing on normative questions about the proper role of state and market, government and industry, public and private modes of managing innovation—issues that in the United States are often themes in electoral politics. Moreover, the year 2000 was a US presidential election year, and the presumptive nominee of the Democratic Party, Clinton's vice president, Al Gore, was expected to face a close race. The upcoming election intensified the macropolitical significance of the media coverage of the HGP versus Celera competition. In this context, Clinton sought to portray investment in science and technology as something that the federal government does right. Thus, it is not surprising that in his annual State of the Union Address in January 2000 he presented the HGP as an example of appropriate and efficacious government investment: "Later this year, researchers will complete the first draft of the entire human genome, the very blueprint of life. It is important for all our fellow Americans to recognize that federal tax dollars have funded much of this research, and that this and other wise investments in science are leading to a revolution in our ability to detect, treat, and prevent disease."[71]

For some champions of market solutions over government action, a view strongly associated with the Republican Party, Celera's widely publicized speed and innovativeness were an illustration of the superior efficiency of the private sector. The competition thus grew relevant to the upcoming election, intensifying the Clinton administration's interest in concluding the "race" and celebrating a historic achievement. In the spring, secret negotiations led to the resolution that the White House wanted: a joint media event involving President Clinton, Prime Minister Tony Blair, NHGRI director Francis Collins, and Craig Venter of Celera. The timing, June 26, 2000, which HGP scientists regarded as "arbitrary" or "contrived," was determined by the availability of an opening in Clinton and Blair's calendars. Clinton spoke at the White House, and Blair participated from London via satellite. Both spoke of the achievements and promises of the genome project in revolutionary terms. Francis Collins and Craig Venter also spoke. CNN broadcast the event live.

Clinton's speech eloquently built an encompassing "we," a broad collective tied together by science, progress, the sacred, a common humanity, and democratic values, and he framed the genome project as an international achievement tightly linked to the imaginary of a single, global world united in the pursuit of shared values. He also linked the sequencing of the genome to the well-institutionalized US imaginary we might call "America the Innovator"—a collective understanding of technological prowess and adventurous scientific exploration as a national trait.[72] Through a reference to the Lewis and Clark expedition that journeyed across the continent at the outset of the nineteenth century, he linked the mapping and sequencing of the genome to the national imaginary of the American frontier:

Nearly two centuries ago, in this room, on this floor, Thomas Jefferson and a trusted aide spread out a magnificent map. ... It was a map that defined the contours and forever expanded the frontiers of our continent and our imagination. Today the world is joining us here in the East Room to behold a map of even greater significance. We are here to celebrate the completion of the first survey of the entire human genome. Without a doubt, this is the most important, most wondrous map ever produced by human kind.[73]

Clinton celebrated the "brilliant and painstaking work from all over the world" that had contributed to this accomplishment, saying that "we have pooled the combined wisdom of biology, chemistry, physicists, engineering, mathematics and computer science, tapped the great strengths and insights of the public and private sectors." Regarding health, he said that genome science "will revolutionize the diagnosis, prevention and treatment of most, if not all, human diseases." And he linked the "stunning and humbling achievement" to the sacred: "Today we are learning the language in which God created life."[74]

Presenting this "historic achievement" as only a starting point, Clinton outlined three future horizons. First, "we will complete a virtually error-free final draft of the human genome before the 50th anniversary of the discovery of the double helix, less than three years from now"—a clear reference to the HGP's commitment to producing a high-quality, finished sequence. Second, "we must sort through this trove of genomic data to identify every human gene," discover their functions, and "rapidly convert that knowledge into treatments." And third, we must address the "ethical, moral and spiritual dimensions of the power we now possess":

We must not shrink from exploring that far frontier of science. But as we consider how to use new discovery, we must also not retreat from our oldest and most cherished human values. We must ensure that new genome science and its benefits will be directed toward making life better for all citizens of the world, never just a privileged few. As we unlock the secrets of the human genome, we must work simultaneously to ensure that new discoveries never pry open the doors of privacy. And we must guarantee that genetic information cannot be used to stigmatize or discriminate against any individual or group.

One of the great truths to emerge from genome science, Clinton continued, is that, "in genetic terms, all human beings, regardless of race, are more than 99.9 percent the same. What that means is that modern science has confirmed what we first learned from ancient faiths: The most important fact of life on this Earth is our common humanity. My greatest wish on this day for the ages is that this incandescent truth will always guide our actions as we continue to march forth in this, the greatest age of discovery ever known."[75]

Clinton's speech represents a convenient marker not only for the "completion" of the HGP but also for the consummation of the transformation of the vanguard vision advanced by the HGP leadership at the outset of the project into a broader sociotechnical imaginary, publicly performed by the head of state and integrated into the historical self-consciousness of the imagined community of Americans. This broader "genomics revolution" imaginary presented the sequencing of the human genome as a world-changing triumph with tremendous potential to better the human condition. The speech also expressed a sociotechnical imaginary, embedded in the HGP's ELSI program, of how to manage the social and ethical concerns associated with emerging science and technology. This imaginary is grounded, on the one hand, in a postulated separation of *science and technology* from its ultimate *applications and use* and, on the other, in an ontological distinction between the *use* and the *misuse* of genomic knowledge, treating this boundary as one that can be discerned and appropriately managed.[76] In this manner, Clinton's speech publicly performed an imaginary in which transformative scientific change is susceptible to governance by a collective "we" that shares the fundamental normative principles of a common humanity. Like the vanguard at the outset of the HGP, he promised an orderly and beneficial revolution—one that will preserve cherished human values in the midst of dramatic change.

* * *

Two more official celebrations of the completion of the Human Genome Project took place: the next in February 2001, when the HGP and Celera simultaneously published separate papers. *Nature* published the HGP's draft, and *Science* published Celera's WGS assembly, with both journals devoting most of their space to the sequencing of the human genome.[77] However, in contrast to the HGP, which made all of its data available without restriction via GenBank and the other databases, Celera made the entirety of its data available only to its subscribers. Nonsubscribers were allowed access free of charge to any small segment of data that interested them, but they could not simply download sequence information without restriction.

Celera's data-access restrictions generated controversy. In particular, some HGP scientists sharply criticized *Science* for publishing Celera's paper without requiring that the data be fully released. Although the company's policy was more permissive in granting nonsubscriber access than are many subscriber-based business models in commercial media, it contradicted a key principle of the regime of scientific publication: namely, that the data underlying a paper be made available to qualified scientists after

publication. The debate over *Science's* decision eventually led to the formation of a National Academy of Sciences committee to review questions
about the publication of such data, and the resulting report in 2003 reaffirmed the principle that journals should require the data and materials
underlying published works to be made available to other scientists.[78]

* * *

At Cold Spring Harbor in May 2001, the mood seemed to me unusually
reflective. The scientists gathered had been part of a human achievement
that they expected to be remembered for decades, perhaps centuries, to
come. They had built new kinds of laboratories, consolidated a field, completed an ambitious project, and self-consciously launched a revolution in
biological research. The race with Celera had been exhausting, and tensions
among and within HGP laboratories had created some resentments, but
balancing these feelings was a deeper sense of accomplishment as well as
excitement about the future. To be sure, much work remained to produce
the finished, high-quality sequence that the HGP had promised to deliver,
and the task of understanding the genome had barely begun. The unexpectedly small number of human genes pointed to new complexities. Some
3 billion letters of code awaited analysis, and the work of finding meaning
in biomolecular information was expected to occupy biologists for much of
the twenty-first century. Indeed, HGP leaders immediately began referring
to the sequencing of the genome as merely the "end of the beginning" of
the genomic revolution.[79]

Few of the Cold Spring Harbor attendees in 2001 seemed to doubt that
the entire sequence would soon be completed (setting aside unsequenceable regions). In fact, the level of confidence was sufficiently high for HGP
leaders to plan to declare the project "done" on a date specifically chosen
for its symbolic significance. The HGP had already decided to pronounce
the human genome "finished" in April 2003, the fiftieth anniversary of the
publication of the famous Watson and Crick paper on the chemical structure of DNA. In light of some unknowns, Kevin explained, "we may have to
change the definition of finished" a bit to meet that deadline, but this
would be a tiny adjustment. There was no "sign of any significant deteriora-
tion of the standards of what we mean by 'finished.'"[80]

As planned, the HGP marked the completion of its final sequence in
April 2003 with the publication of a paper in *Nature*, the journal where
Watson and Crick's paper had appeared in 1953. With this announcement,

the genome project was successfully completed for a third and final time. The NHGRI was thus able to convincingly claim that it had succeeded in achieving its stated goals ahead of schedule and under budget, defining the Human Genome Project to be an unqualified success. Celera did not become "the future leading provider of genomics knowledge," as one enthusiastic investment analyst had predicted in 1999.[81] Controversy about Celera's achievements continued. About 60 percent of the reads used in the company's famous paper in *Science* in February 2001 consisted of HGP data that the company had downloaded, a fact that raised questions about the extent to which the Celera WGS represented a truly independent assembly of the genome.[82] The company was able to sell high-priced subscriptions to its data for a time. But with essentially the same data available for free from GenBank, the market for Celera's sequence dried up, and the company unceremoniously fired Venter and changed its focus to diagnostics.[83]

A Symbolic Resource

As the HGP drew to a close, the notion of a project finished ahead of schedule and under budget was of critical symbolic and political importance. Well before the *Nature* and *Science* publications in February 2001, science policy makers began considering how the sort of regime that governed the HGP could be applied to other research domains better accomplished through "large-scale projects" than through the investigator-initiated grants of "ordinary biology." In an interview in July 2000, held shortly after the Clinton/Blair announcement, JIM, a manager at the NHGRI, explained:

One of the reasons that the genome project as we knew it was as successful as it was is because it was able to identify a specific data set that was worth collecting, was best collected in this targeted, centralized, quantifiable, so forth, manner. So then to me it seemed obvious that the next question was, "What's the next data set that can be approached like that?" I think it was pretty clear coming out of the [recent] planning process is that it's not obvious. There are a few obvious candidates—all of the full-length cDNAs, the mouse sequence, the rat sequence—but that is not as conceptually as big a step, they don't provide the challenge that the human sequence did 10 or 12 years ago.[84]

The NHGRI, aided by its broader networks of HGP leaders, had no trouble identifying a number of other data sets to produce via large-scale science projects, such as creating a map of human genomic variation. Genome scientists and other members of sociotechnical vanguards immediately began promoting "large-scale biology" in articles and science policy reports.[85] The success of the HGP, rendered into a visible historical fact by the public performances and celebratory discourse, thus became a compelling exemplar of a new mode of doing biological science. Like the Manhattan Project and the Apollo moonshot before it, the HGP was incorporated into the US national sociotechnical imaginary of "America the Innovator"—even as it was also presented as the achievement of a common, global humanity. In the years since its completion, the Human Genome Project has come to serve as an important part of the symbolic repertoire that members of sociotechnical vanguards use when envisioning transformative scientific change and regimes for governing it.

8 Conclusion

Despite its symbolic importance, the completion of the Human Genome Project by no means marked the end of the genomics revolution. Since the turn of the century, genomics has continued to develop, and genome technology has grown much more powerful. Researchers have sequenced the genomes of many species and, to better understand human diversity, many human individuals. New "postgenomic" vanguards have pursued research agendas under such rubrics as systems biology, synthetic biology, and epigenetics.[1] Researchers have incorporated genomic methods into nearly every field of the life sciences. New sequencing technologies have emerged, the cost of sequencing has declined exponentially (dropping by 2015 to roughly $1,000 per human genome), and the volume of data in GenBank has continued its relentless increase. Biomedical applications have so far been relatively modest, although researchers have made promising advances and identified a wide range of genetic variants that influence the risk of common diseases. Genome sequencing has also begun to be introduced into clinical practice, most notably in cancer treatment, with some promising results. Scientists predict that as much larger numbers of human genomes are subjected to comparative analysis, the complex networks implicated in most human disease will be unraveled. One genomics company, BGI, formerly known as the Beijing Genomics Institute, plans to sequence the genomes of more than a million humans, plants, and animals. The speed of change makes it likely that many of today's impressive achievements will look routine in only a few years. Genomics leaders such as Francis Collins, who was made the director of the NIH in 2009, are confident that the genomics revolution is still at an early stage, believing that "the best is yet to come."[2]

As this revolution has continued to unfold, new knowledge-control regimes and problems of governance have rapidly taken shape. Genomics has become an increasingly global enterprise, and a variety of visions of how to create and capture control over emerging knowledge have been advanced. The vast quantities of genomic data now being produced are being selectively enclosed and disclosed under many different regimes, and genomic information is increasingly being linked to other sources of "big data." Business models in personal genomics have reconstituted agents, both corporate and individual, in ways that endow them with new capacities and incapacities over the control of genomic information. Questions about how to govern genomic technology are being raised at an accelerating pace, as new controversies have broken out about control over predictive testing, ownership of clinical genomic data, and the wise use of new and powerful gene editing technologies such as CRISPR. Such developments, which I can barely hint at here, have only increased the need for theoretical frameworks equipped to analyze the processes— within the laboratory and beyond it—that are reordering the worlds in which we live.

This book has elaborated an analytic perspective for investigating the reconfiguration of knowledge and control during periods of transformative scientific change. Like legal institutions and the law itself, what I have called knowledge-control regimes are embedded in society and ongoing practice, operate on many scales, and are used to put into effect many specific forms of control. Analyzing them requires a theoretical vocabulary suited to comparing regimes, whether nascent or long standing, and to examining their emergence, transformation, and interaction. Such concepts as control relationships and governing frames are intended to facilitate examination of how these regimes give structure to entities and relationships, without stripping actors of agency or implying that actual practice neatly tracks the lines of action that governing frames promote.

With this focus on knowledge-control regimes, I have investigated a number of questions about the coproduction of knowledge and control during the HGP: As distinctively "genomic" forms of knowledge and technology took shape, how were the lawlike regimes that governed biological research adjusted? What kinds of changes did the genomics vanguard try to implement, and how did competing factions struggle to advance different visions of knowledge and control? How did actors attempt to use, resist, and reconfigure existing regimes and practices, and in what ways did they seek to reallocate rights, duties, privileges, and powers? And as new forms of knowledge took shape, which regimes and practices were reconfigured, which were not, and why? Along the way, I have examined the formation of new regimes at the outset of the HGP, the struggles over a set of important knowledge objects, the process of regime change and interaction at points of jurisdictional contact, and the production of narratives of justification for specific forms of order. In light of the complexity of these interactions, one would be ill advised to present sweeping generalizations about them. Nevertheless, it is useful to sketch out several tentative observations that this analysis of genomics during the HGP suggests.

Knowledge and Control

The epistemic problem of securing knowledge and the sociopolitical one of securing control are deeply, even inseparably, intertwined. Knowledge objects inevitably take shape within the jurisdictions of specific knowledge-control regimes, and from the moment of their first emergence knowledge objects are thoroughly embedded in control relationships. Yet more than mere embedding is involved. The ethnographic approach used here has shown that control relationships do not merely *surround* knowledge objects but are also actively *built into* them—for instance, when the perimeter of the laboratory is inscribed into the content of scientific papers or packaged into the specific form of circulating biomaterials. Constructing knowledge objects, choreographing their transfer, and managing the perimeter of the laboratory are part and parcel of a single process. Control relationships are also implicated in establishing *who* an agent is—whether we are talking about a well-institutionalized type, such as the head of a molecular biology laboratory, or an emerging type, such as the genome center at the beginning of the 1990s. Objects, jurisdictions, and agents take shape along with knowledge in an ongoing process of coproduction.

From this perspective, the laboratory appears not only as a site for the manufacture of knowledge or solely as an actor in some form of economic or quasi-economic exchange but also as sociotechnical machinery used to constitute control relationships and endow objects and agents with specific capacities and incapacities. The same can be said for the sociotechnical machinery associated with many other sites implicated in coproducing knowledge and control, including the funding agency, the biotechnology company, the journal, and the patent office, not to mention the transorganizational linkages among them. Actors' capabilities and even their identities may be reshaped when they become entangled in new control relationships, as illustrated by the example of the university researcher who is drawn into the machinery for colonizing knowledge that I called the HGS nexus—a transorganizational network of control relationships linking university scientist to biotechnology startup to multinational pharmaceutical company. In short, this examination of genome research has unpacked how lawlike regimes of varying degrees of formality and stability operate in practice and how these regimes are reconfigured as new knowledge and technology take shape, as the capacities of objects and agents undergo change, and as jurisdictions and control relationships are renegotiated.

Examining genome research has also documented how lawlike regimes sometimes fail to achieve complete and unproblematic control over the agents, knowledge, and spaces that they regulate. Friction is common, and actors use a wide range of techniques to resist efforts to structure their capabilities and duties. The perimeter of the laboratory is always vulnerable to being breached. Jurisdictional disputes may break the governing frames of the ad hoc regimes that sustain collaborations. Actors may evade contractual controls, exploit the limits of regulatory reach, or redefine situations in ways that expand their options. But even given the inevitable limits of all regimes to enforce order, specific changes in knowledge-control regimes may profoundly alter allocations of cultural authority, wealth, and power.

The Dynamics of Change

The account of transformative change developed here centered on the activities of sociotechnical vanguards. In this account, the member of a sociotechnical vanguard looks not like the over-socialized agent of Mertonian sociology, nor like the under-socialized agent of neoclassical economics, but like an active interpreter of ambiguous rules and emerging knowledge—a bit like a constitutional lawyer—who engages with and at times reformulates extant categories, reasoning, and regimes of governance.

As the members of the genomics vanguard, a loosely bounded and dynamically changing collective, worked to revolutionize biology and reorder life, they pursued visions that overlapped only partially and at times were fundamentally incompatible. Nevertheless, even when competing factions developed, as they did in many ways documented in the previous chapters, and even as they struggled with one another, their visions continued to share many elements, and the factions still contributed to the overall movement that built genomics. As part and parcel of this process, the members of the vanguard worked not only to articulate visions but to actively prototype, build, and constitute knowledge and regimes intended— quite literally—to *realize* their goals. These efforts prominently included trying to enroll supporters, gather resources, and secure knowledge and control. They attempted to accomplish these goals not by creating change entirely de novo but by remaking and reconfiguring the material, discursive, and social resources available in the worlds in which they were embedded. The resulting picture of transformative scientific change thus emphasizes a coproduction process involving the activities of vanguards; their internecine struggles and interactions with other groups; and their reliance on ready-to-hand resources—including existing knowledge, regimes, practices, imaginaries, and other cultural forms.

This examination of genomics during the HGP sheds light on the challenges that vanguards encounter when attempting to launch new modes of securing knowledge and control, especially in cases that require significant changes in established lawlike regimes, cultures, and practices. Building a truly "new" regime, for example, entails persuading a variety of actors to accept burdens and entitlements that reorder established frameworks, practices, and perhaps self-conceptions and identities. The RLS provides an illustration. This regime emerged in a single laboratory that envisioned and built into material form a truly novel way to constitute genome mapping. But its ambitious goal—of integrating a network of laboratories into a collective knowledge-producing machine that would avoid duplication of effort and rapidly accumulate data—conflicted with the institutionalized practices of molecular genetics, and so the RLS became a road not taken. This outcome is precisely what one might expect in light of the intense individualism embedded in the culture and practices of molecular biology. But the failure of the RLS vision in no way implies that this form of scientific life was in principle impossible. There is no reason to naturalize the culture and practices of molecular biology in the 1990s, assuming they were functionally necessary for efficient knowledge production. Nor is there any reason to reject out of hand the possibility of building knowledge-control regimes that constitute tightly integrated collectives, held together by shared goals, materials, and practices, that seek to optimize overall achievements through rapid circulation of data and materials.[3] My argument is not that regimes such as the RLS cannot in principle be constructed but that this vision proved impossible to realize in a scientific culture that imagined itself as a population of autonomous, atomized individuals engaged in what some scientists, especially in human genetics, seemed to regard as nearly Hobbesian competition.

New regimes proved easier to constitute when they did not attempt to change the control relationships operative in established jurisdictions. Rather than attempting to impose new burdens and entitlements on molecular biology laboratories, the US genome program distinguished the HGP from "ordinary biology," constituting a governing frame focused on controlling the new agents it sought to bring into being (genome centers) without disrupting the existing ones (ordinary laboratories)—at least in the short term. Significant change in ordinary laboratories soon did begin to take place, but it stemmed from the achievements of the new regime rather than being a precondition of the regime's implementation.[4] Later, when the HGP developed the Bermuda Principles, new and unprecedented duties were once again imposed only on the new, leading centers of genome research: the large-scale sequencers.

The vanguard that launched GenBank was also able to begin its project without changing established lawlike regimes and cultures. The staff-driven collection regime could begin without intruding into the jurisdictions of journals or the laboratories that published scientific papers containing sequences. Unconstrained by cross-jurisdictional entanglements, the Los Alamos group built a database that soon became indispensable. Intractable problems developed quickly, however, as staff-driven collection and voluntary direct submission failed to keep pace with sequencing. But to mandate direct submission—by compelling laboratories to send their sequences to the databases upon publication of a paper—required reconfiguring control relationships outside of GenBank's jurisdiction. No longer could the journals and the databases merely *coexist*; they would have to actively *cooperate*. The success of the negotiations that accomplished this transition reflected the growing importance of sequence data, the support of the funding agencies, and, not least, the fact that the supporters of mandating submission could plausibly claim to be building on a precedent: the principle of the journal regime that authors must provide qualified scientists with the data underlying published work.

The debate about the control relationships for managing what I called UJAD data—sequence information that was both unpublished in journals and available in databases—further illuminates the difficulties of reordering well-established regimes. The challenge of regulating these data, which fit poorly into the tidy ontology of the scientific publication regime, arose after some researchers downloaded entire data sets produced by the large sequencing laboratories and then published papers that preempted publications that the large sequencers had planned. Genome researchers struggled with the question of what rights, if any, large-scale sequencers retained in the data sets that they had produced. Any "right of first publication" needed to be carefully bounded or it would represent a significant departure from established practices. In this case, the sense that the sequencers had a legitimate complaint was sufficiently strong to allow a loose regime, grounded in the language of "etiquette," not rights, and supported by a distributed system of informal responsibility, to emerge at the Fort Lauderdale meeting.

Taken together, these examples suggest that, other things equal (as they rarely are), we might expect that substantial changes in lawlike regimes would be more likely to materialize to the extent that those changes (1) are consistent with prevailing cultural forms and extant regimes; (2) do not increase the burdens imposed on agents who can influence the success of the regime; and/or (3) can be implemented without requiring negotiations at points of regime contact. We also might expect that concentrations of expertise, wealth, political clout, and configuration power would ensure that some actors are especially well positioned to promote and resist changes in lawlike regimes—as Merck did when it decided to challenge the SmithKline-backed HGS nexus by funding an effort to produce a counter-collection of ESTs for public release, or as the funding agencies did when they backed the Bermuda Principles and later the Fort Lauderdale Agreement.

My examination of knowledge and control during the HGP also calls attention to the importance of interactions in which the ontologies of science and lawlike regimes (and sometimes even formal law) change *together*, producing new objects that are irreducible hybrids of knowledge and control. One dramatic case of such ontological change was the ongoing and contested transformation of cDNA-related objects, especially after they were refashioned not as incomplete results but as tools for rapidly discovering genes and capturing control over substantial parts of the human genome. The vision of cDNA patents as a means to win control of a treasure trove of intellectual property was short lived, but as this example highlights, even changes that are not durable sometimes have notable effects. In this case, the NIH patent applications provoked the United Kingdom to freeze access to its cDNA sequences and file defensive patent applications in retaliation. The vision of ESTs as a means to capture intellectual property also ignited the enthusiasm of venture capitalists, setting the stage for the launch of a wave of companies aimed at transforming genomic information into biocapital.

In the longer run, however, the ontological changes that seem to have had the most lasting effect on the rise of genomics are subtler ones, such as the distinction—so critical to getting the HGP off the ground—between "genome research" and "ordinary biology." This distinction provided the rationale for establishing some sort of human genome project in the first place as well as for constituting new regimes for the express purpose of governing the project. Once stabilized, this distinction could be deployed to justify a variety of accountability measures, from the STS standard to the Bermuda Principles. This distinction was eventually broadened and extended in the Fort Lauderdale Agreement's concept of the CRP.

* * *

Sociotechnical vanguards seek to make futures, but they cannot make them simply as they please; they do not make them under self-selected circumstances but rather use cultural resources and practices already given and transmitted from the past. The ready-to-hand resources and prevailing sociotechnical imaginaries of the late twentieth century structured the terrain on which the leaders of the genomics vanguard developed their visions and exerted their agency. These actors were embedded in complex sociotechnical worlds replete with the contradictory tendencies and normative disagreements found in any social order. As they worked to develop and realize their goals, they drew on the conceptual and institutional furnishings of the sociotechnical orders in which they were situated. Well-institutionalized regimes, such as that of the scientific journal, and well-capitalized institutional machinery, such as the venture capital system and the policies and practices underlying "startup culture," proved very influential. So did imaginaries of scientific openness, biomedical improvements of health, and science as a source of wealth and power. As the case of genomics during the HGP suggests, the imaginaries, regimes, and resources that vanguards use to advance their visions do not do so in a "neutral" way, but differentially nurture particular modes of constituting knowledge and allocating control. The broad tendency is for the most radical visions to fail to be fully institutionalized or to be adjusted into forms that do not require abrupt departures from dominant orders. Grand visions of technology as a source of utopian social change tend to underappreciate both the resilience and the obduracy of extant orders.

Understanding the coproduction of knowledge and social order is a significant challenge for the social sciences today. This study investigated the ways in which the production of knowledge and the production of control are intertwined, even in spaces—such as laboratories and research programs—that often seem institutionally and temporally distant from applications and societal impact. Analyzing knowledge-control regimes and practices helps bring the political dimensions of these spaces into fuller view—a move with ramifications for understanding the social dimensions of a wide range of technoscientific fields.

The centrality of science and technology to the politics of the twenty-first century, at scales ranging from the geopolitical to the interpersonal, is no longer open to serious question. This account of the rise of genomics suggests that those who wish to understand the dynamics of power in contemporary societies cannot afford to ignore knowledge-control regimes and the informal practices through which knowledge and control simultaneously take shape. For if the promoters of emerging technoscience not only create knowledge but also constitute agents and relationships, if they not only map genomes but also redraw jurisdictions, if they not only make tools but also configure users with specific (in)capacities, if they not only produce information but also allocate power over the direction of sociotechnical change, then we cannot safely assume that technological innovation is simply a rising tide that raises all boats. If knowledge and control are coproduced, then understanding contemporary societies requires recognizing the vanguards that champion scientific revolutions as the political actors that they are.

Appendix: Fieldwork and the Control of Knowledge

This study could not have been completed without the generosity of my interlocutors, who took the time to explain and discuss their fascinating work and worlds with me, often on multiple occasions. They allowed me to visit their laboratories, attend their meetings, and ask them a huge variety of questions. I am enormously grateful to them for the gift of their precious time and for their willingness to contribute to the work of a researcher from such a different scholarly tradition. It was a privilege to interact with them, and I remain in their debt.

Questions about how social scientists acquire and manage information arise frequently in discussion of research methods and ethics, and in the study of knowledge and control these issues seem especially salient. It is therefore worth taking a reflexive look at my own research methods and confidentiality practices, and in this appendix I give a sense of the knowledge-control regime implicated in the construction and reporting of the findings presented here. My research practices varied with different modes of data production, so here I look at interviews, participant observation, and written documents separately.

Research Interviews

I conducted interviews with genome scientists, including directors of genome research centers; program officers in funding organizations; members of advisory committees; researchers in genome laboratories (including senior scientists, postdoctoral associates, graduate students, and technicians); and company officials. Many of these people were interviewed more than once, sometimes years apart, and many of these people changed roles or organizations over the course of the study. Some of my interlocutors were critics of aspects of the HGP, and some were scientists in overlapping/ adjacent fields, such as human genetics. The bulk of these interviews were conducted in the United States, but genome scientists were also interviewed in the United Kingdom, Germany and France. Interviews, which typically were about one hour in length, with almost all ranging from 30 minutes to two hours, were audiorecorded and transcribed, and a variety of published and unpublished documents were also collected. Interviews were often accompanied by laboratory tours. Most of these interviews were conducted in academic settings or national laboratories, but some also took place in government offices, nonprofit institutes, and (after genome companies emerged in the mid-1990s) commercial firms. In all, 190 formal interviews were conducted, audiorecorded, and transcribed.

From the perspective that underlies this study, a research interview is simultaneously a face-to-face interaction featuring the selective revelation and concealment of information, a bounded "space" containing knowledge, and a source of information that is selectively reproduced and transformed as the research process produces new knowledge objects. Unlike most face-to-face conversations, these encounters are specially requested and scheduled for the announced purpose of conducting an interaction framed in the precast roles of interviewer and interviewee about a specific topic—in this case, the topic of genome research. In these interactions, I faced dilemmas of self-presentation familiar to field researchers, some of which may have been amplified by the challenges of "studying up" and of doing research in a complex and fast-moving technical arena that I could not hope to master. These dilemmas included striking the right balance between being knowledgeable enough to be credible and sufficiently uninformed to ask the probing and sometimes naive questions needed to uncover background assumptions and mundane routines. In addition, most of my interlocutors were very busy, so I had to continually make strategic judgments about how to use limited interviewing time. (I typically asked for, and usually received, an hour to an hour and a half.) My efforts to manage these interactions met with varying success, but overall I learned a great deal from interviewing.

The interview excerpt given here, chosen because it is unremarkable, illustrates the typical question–answer structure that the interviews tended to follow. This strip of talk, which began some 3,600 words into a

21,000-word interview, pertains to the problem of "background noise" in detecting "signals" from fluorescently labeled DNA molecules:

SH: [The noise is] enough so that at this level you start to run into a significant limiting kind of problem?

Jason: Yes. Now, the extent to which this happens is a thing that we don't have extremely good control over. It varies from prep to prep. It varies from technician to technician who makes the preparation. Some people are very, very stringent about keeping their reagents ultraclean and doing their washes properly and adhering assiduously to the protocols that have been optimized by people like PETER POSTDOC. Other people are somewhat sloppier. They don't always get the best results. I think in fairness it has to be said also that it varies from preparation to preparation in ways that no one is really sure of why it does.

SH: The reagents that you [use] may have variance in them as well.

Jason: They may well. That's absolutely right. Fairly recently, ultrapurified reagents have become available, like avadin and streptavadin that has been purified specifically for its performance in florescence in situ made a significant difference in the quality of the images that you get.[1]

Like most of my formal, audiorecorded interviews, this interview took place out of earshot of others, so the immediate audience for each participant thus consisted of one other person.[2] Our interaction thus took place in a "space" bounded in four interconnected ways: discursively (the topical focus on genome science), temporally (May 1990), physically (a university building), and socially (SH and JASON—a pseudonym, as indicated by the small caps). Such boundaries may be more or less tightly circumscribed, but clearly no conversation takes place in a hermetically sealed world, entirely cut off from everything and everyone, from which information can never be deliberately revealed or inadvertently transmitted. The participants in all conversations may retain records or memories, albeit imperfect ones, that provide material that actors sometimes process and package into accounts presented to future audiences. In principle, these repackaged accounts may flow through channels as carefully targeted as a soft whisper into a well-chosen ear or as indiscriminant as a CNN broadcast.

Because these interviews were intended to underwrite publications, they were designed to produce records, and these records fell into eight main categories, not all of which are relevant to all of the interviews: (1) audiorecordings, (2) verbatim transcripts of those recordings, (3) notes scribbled during the interview, (4) correspondence pertaining to the interview, (5) materials gathered and notes prepared beforehand, (6) documents that interviewees gave me during or after the interview, (7) diagrams or pictures that interviewees drew to explicate a point, and (8) comments on interviews that I produced after the fact. When quoting from interviews in this book, I present verbatim excerpts from the transcripts. In displaying quotations, ellipses indicate elided material, usually omitted because I deemed it redundant or extraneous. Brackets are occasionally used to insert a clarification.

The prospect of publications and other retellings of information revealed during interviews inevitably raises the question of precisely what information might be transmitted to whom in the future. In social research, the possibility that such information might spread to future audiences in ways detrimental to the interests of the participants (an ever-present one in ordinary conversation) is regulated via a knowledge-control regime that is established for the protection of human subjects and intended to instantiate a distinction between retellable and not-retellable parts of what transpires. In the case of the research I conducted, the people interviewed were assured that their identities would not be revealed.

A consent form described the ground rules for my revelations of information. Consent forms, of course, serve multiple purposes: protecting human subjects by specifying their rights; protecting researchers and the institutions that employ them; helping to "sell" participation in the project to potential participants by assuaging anxieties; and so forth.[3] The form conveyed a number of messages. It professed a lack of interest in individuals per se: "Unlike journalists, who are interested in stories about individuals, sociologists are interested in social structures and processes. There is a tradition in most sociological field research of maintaining the anonymity of research subjects." And it stated that the names of people interviewed would not be revealed: "You will not be identified by name as a source of interview material without your written permission." My interlocutors were promised that the unedited records from the interviews would not be distributed and were told that excerpts of the transcripts may be quoted or paraphrased when reporting the findings of this research.

The consent form also described how publications would repackage the identity of interview subjects to provide (some) context to audiences without revealing specific identities: "The source of interview materials will be described in general terms (e.g., a source might be identified as a 'scientist at a major genome sequencing center,' a 'human geneticist at an eastern university,' or a 'technician who works with microarray technology'). In some cases, pseudonyms may be used." It also acknowledged the limits of such repackaging: "In some cases, it is likely that the work that a particular laboratory does will be sufficiently unique that knowledgeable people could make accurate guesses about its identity. However, when particularly sensitive issues are involved, Professor Hilgartner will take extra steps to help disguise identities." The consent form did not describe these "extra steps." However, in this book they include the omission of identifying details and in a few cases use of the occasional "red herring," such as a misleading geographic location. Indeed, the pseudonyms that involve place-names in the United States, such as WESTERN GENOME CENTER, may be misleading. In some cases, the same person has been assigned more than one pseudonym.

I take human subjects protections seriously and am willing to pay the epistemic price that doing so imposes. In some cases, the information-control practices employed significantly reduce the strength of the arguments I can make or weaken the empirical material supporting them. The seriousness with which my interlocutors took the informed-consent process varied. Some carefully read the consent form and handled their copy in a manner suggesting that they would carefully file it. Others viewed the consent process as a bureaucratic formality, barely scanning the form and in a few cases even tossing their copy away in my presence. When presented with the consent form, one European scientist scoffed that the United States would soon sink into the sea under the sheer weight of its innumerable lawyers.

In the typical biomedical study, concerns about confidentiality stem from the fact that researchers may collect detailed information about bodily pathology or stigmatized behavior. I did not collect data on bodily pathology, but concerns about stigmatized conduct, personal privacy, reputation, and the presentation of self clearly arose in my research. Some people seemed particularly concerned about revealing negative opinions or information about colleagues, whereas others seemed worried about overclaiming their own achievements, and still others regarded ongoing negotiations with potential funders or collaborators as especially problematic. Another layer of apprehensions about confidentiality, more specific to the world of emerging technoscience, pertained to the possible transmission of ideas or unpublished findings to potential competitors.

Dylan: Off the record, if I can talk about confidential stuff? Can I do that?
SH: Yes.
Dylan: So this is, we've just had a major success in my lab. That's not public yet because we are writing the paper, but, and the public database has made it a lot faster than it would have been otherwise. We cloned a disease gene.
SH: That's exciting.
Dylan: I'm really really, I've seen numerous—
SH: Is it in the publication pipeline now?
Dylan: No, I'm writing it now, so that's why I can't, OK, but if you want to talk to me about this in about a month, I'll be happy to tell you all about it.[4]

Despite abundant evidence of such worries, which my interlocutors sometimes made quite explicit, I was often impressed with their willingness to discuss potentially sensitive aspects of their work, in both formal, taped interviews and informal encounters. This is not to say that I never wondered about what was being omitted by those who presented themselves as open. Nor is it to say that my subjects did not at times explicitly wall off certain topics. However, for an analysis concerned with the control of knowledge, such maneuvers are quite revealing, as are the reasons given (or not given).

In their own work, genome researchers are uniformly enthusiastic about the advantages of "automated data capture" in terms of accuracy and completeness, and they perceived audiorecording to exhibit these advantages. Only a minority too tiny to bear mentioning declined to be recorded, and, as many field researchers report, it often seemed to me as if most interviewees quickly "forgot" that the tape was rolling. Even so, on a few occasions people deemed certain matters too sensitive to commit to a recorded form. The interview excerpt here, in which a US government employee offers to give an account of how a specific organization came into being, is a case in point. The verbatim transcript reads:

Jacob: There's an amazing story about how they were created, which, if you turn off your tape recorder, I'll tell you.
SH: OK.

[Tape clicks off. He tells the story. Tape clicks on.]

SH: The real problem being?
Jacob: Well, I think the problem is you are buying technical resources that you don't really understand.[5]

Requests to turn off the recording device occurred infrequently. A more common request—though still a relatively rare one—involved people explicitly identifying certain details that they did not want retold, at least in connection with "sensitive" topics.

SH: I suspected they were spending [those funds] down because it was such a large amount. I didn't imagine they could raise ... an amount that would endow the amount of money they're spending annually.
Clark: That's right.
SH: But that's interesting to confirm.
Clark: Needless to say, this is all confidential stuff because we still have our [dealings] with RESEARCH GROUP.[6]

Sometimes the consent form would itself prompt a discussion of the difficulty of concealing identities, as the following example illustrates:

[Consent form is signed. Tape recorder clicks on.]

SH: So you just told me that you're the only person in the United States who does precisely what you do.
Nicole: The technology is limited to this particular lab now, so other labs are trying it, but—so if there was a discussion of the technology or what was being done, it would be very clear that it was me who was the person who was involved in this part of the interview.
SH: So that if there was something sensitive, you'd like it separate from a description of the particular technology.
Nicole: Right.[7]

In formal, audiorecorded interviews at commercial firms, the need to protect proprietary information was sometimes mentioned, and on occasion I was told that certain things could be discussed only if I signed a nondisclosure agreement. For example, one company official, TIM, suggested at the outset of an interview that possibly some of my questions would be "better handled by revealing tangentially some confidential information. If there are [such questions], then I would be very happy just to give you one of our confidentiality forms to sign."[8] In this case, I did not sign such an agreement.

At times, my interlocutors turned the questioning on me, seeking information on my views about the genome research community and even about specific laboratories or people—as in the next excerpt, in which a researcher and I dance around her efforts to learn what I thought about other sites.

Kasia: Because I know from having tried you to get to answer things before, that you won't do that much.
SH: Oh, you mean I—

Kasia: I'm curious about—
SH: Like what my conclusions are or something?
Kasia: What your perceptions are of the kinds of changes that you see in the last
SH: [overlapping] Here?
Kasia: [overlapping] few years.
SH: I can talk more about here.
Kasia: I'd be interested.[9]

I told her a few of my relatively unthreatening observations about the laboratory, which was easy because I regarded it as a rather impressive place. These comments led to a back and forth of about 400 words. But she seemed to want to know what my perceptions were of both what was going on "here" and what was going on elsewhere.

SH: You asked me to give you some of my perceptions, so—
Kasia: I'm very interested.
SH: So those are the ones relevant to right here. I'm a little more, I don't want to start spreading rumors within the community, and that's one of my— So I don't want to talk about specific places really, so I am reluctant to do that.
Kasia: [laughs] I never go to the meetings.
SH: That's just part of the thing. On the other hand, I don't want to be terribly evasive, and if you're interested in where I've been and who I'm looking at, I can tell you some of that, too.
Kasia: I know a bit about the community. I'm interested in it as a community. I'm interested in it for my own personal reasons just because I find it fascinating.[10]

She went on to discuss why in her view the genomics community was fascinating, and then to other topics. The interview never returned to a discussion of specific places or people.

Participant Observation

A second form of data in the book, participant observation in laboratories and at scientific meetings and policy-related meetings of various types, afforded me the opportunity to observe research activities and discussions and to take part in or witness many conversations. I conducted ethnographic observations in major genome laboratories in the United States. I visited most of these laboratories more than once, usually for several days at a time. In one laboratory, I did more extended participant observation over a period of about a year and a half, studying the laboratory through repeated visits and even by participating (part-time) in some experimental work. I also visited laboratories in the United Kingdom,

Germany, and France. Field work at meetings was especially important. I attended parts of ten annual meetings (1989–1995, 1998, 2000, and 2001) on genome research held at Cold Spring Harbor, where the community involved in the HGP gathered each year for five days. In addition, I attended a number of other genome-related meetings, including Human Genome I and II, two science advisory group meetings in Washington, DC, two chromosome-mapping workshops, a US genome program planning meeting, four relevant National Research Council workshops, an international meeting on policies for governing chromosome mapping, and a variety of other conferences and workshops in several countries. My service on NCHGR and NHGRI ELSI grant review panels and as a member of the Committee to Evaluate the Ethical, Legal, and Social Implications Program of the Human Genome Project (1996) provided additional background information, although I do not quote from the confidential discussions that took place in those settings.

During fieldwork, I took extensive notes and engaged in countless verbal encounters with people, conducting conversations—or what field researchers regard as "informal interviews"—that were not audiorecorded but were documented after the fact (as soon as practically possible) in dictated or written field notes. In the field, I made no secret of the fact that I was a researcher studying the genome community, and I tried to make sure, when introducing myself to individuals or small groups, that people knew I was an observer. As many textbooks in field research methods recommend, I fashioned a quick description of my research—what the high-technology world calls an "elevator pitch"—to give people a general sense of what I was doing: "I'm a sociologist who is studying the genome-mapping and sequencing community, and I'm interested in how scientists go about doing their work, how they gather resources, and also patterns of communication and interaction in the genome community. I'm interested in emerging technological systems, and genomics is a very interesting area, both in and of itself and as a model system for understanding emerging science and technology." People usually didn't ask too many questions. That the genome project would attract scholarly attention seemed natural enough to most of them given their perception of its historical significance. When they did have questions, they often asked about my "background in biology" or how I "got into" this area. More than a few struck an amused tone, commenting that I must have some "great stories," remarking that I must have dug up some very "interesting dirt," or joking that the genomics community needed a psychiatrist not a sociologist. I would explain, in a duly serious tone, that my interest was in understanding social patterns and dynamics rather than individuals and would endeavor to deflect as smoothly as possible any inquiries aimed at acquiring information that would betray confidentiality.

I did not use consent forms for these informal conversations, although people would typically hear some version of my elevator speech. Whipping out a consent form repeatedly in such settings as a scientific meeting or a cafeteria table would have been, quite simply, too disruptive of ordinary activity to be a sustainable practice. But consent form or no, this study applies the same principles for protecting identities to these informal encounters as to the formal interviews. The unedited field notes, full of names and identifying facts, are treated as confidential materials; my tales of the field, like the interviews, are reported using pseudonyms and when necessary have been edited to remove identifying details or obscure identities (employing the same "extra steps" used in the case of interviews).

I was often impressed by my interlocutors' willingness to discuss their research and to answer my questions about their world and practices. In general, people appeared to be very forthcoming. Exceptions were relatively rare, and it is worth considering an example. Early in my research, I attended a meeting where I had a chance to catch up with a genome researcher I knew, whom I call Jay, and to trade accounts of our current projects. I told Jay about my plans for the next phase of my research, and he said that he had invented a new technique, described in his grant proposal. I inquired about the technique, but to my surprise, given my previous experience in most of my dealings with this research community and with Jay himself, he wouldn't tell me what his technique was. An account constructed by editing my field notes explores his reasoning.

Jay didn't want to talk about it because, he said, he was in a difficult position. He had to worry about getting his results out, and he was concerned that, "well, there aren't that many ideas out there." Clearly, he was indicating that he was unwilling to reveal the nature of the technique to me because he feared that others might learn about it from me before he had a chance to make it really work. Most people can't follow up all the ideas they have, he said, so what happens is they think of something, and they don't really pursue it. Then later on they hear that someone else is beginning to pursue it or that somebody else has written a really strong argument in a grant proposal for why it is a good idea, and then they say, "That really is a good idea," and they maybe throw some resources at it. Or after someone publishes about it and it becomes a working idea out in the world, they say, "Oh, boy, I should have done something with that. I thought of that two years ago." He clearly wanted to avoid such an experience.

Jay also perceived another risk for people like him whose careers were not secure. Well-known people, like KATZ, publish position papers where they say, "This strategy is a really good way of doing this. This technique might work," and so on, but they haven't actually made it work. Then some poor guy, who may have already been using that technique but was just trying to get the results really in good shape before he went to publish, that guy now has to cite, reference one, "KATZ has proposed this idea," even though the unknown guy thought of it independently and first.

In discussing my own research, Jay also told me that he thought few people would be willing to tell me about their new ideas. Now, obviously, since he had just told me that he didn't want to tell me about his new technique, what he went on to say was easy to view, at least in part, as his justification for that refusal as well as a face-value comment on what he thought other people might think.

One concern Jay raised was that people might wonder: "Hey, I go and I talk to this guy, this sociologist, does that affect my patent claims? Does that constitute publication, and, if so, would that be incompatible with patent protection? If a question arose about the date of the invention, could this guy's notes get subpoenaed, and could he end up saying, 'No, no, no, they hadn't made that technique work in December'?" He said that the people I wanted to interview would also be concerned about how ideas might be spread around, both to competitors and to companies that have a financial stake in things. Since new techniques in this area of science are potentially patentable, and since everything scientists are doing has got commercial possibilities, he said, the potential for spreading ideas is a real concern.

Also, Jay said, people may be very worried about your impact on historical accounts of what went on. He said that if one talks to old scientists, they say, "'Well, so and so got all the credit for that but really what happened was—' and then they tell about some graduate student who really thought up the idea." Given those kinds of difficulties, given those kinds of different versions of the story, people won't want to have somebody who was there in some sort of a semiofficial status, as an outside observer who could give the definitive account of what really happened.

I responded, "Yes, well, people could easily challenge my account."

"Sure, they can say, 'Well, he wasn't really a scientist, he really didn't know exactly what was going on,' or something like that."

"Sure, that's what they would say, right?"

"But even so, there's only one truth about who invented something, discovered something, and they wouldn't like having somebody around to call things into question." Jay recommended that I try to get a legal opinion stating that letting this sociologist visit your laboratory or interview personnel in your laboratory was not going to affect your patent rights.[11]

I did not attempt to get such a legal opinion. In conducting my field-work, I sometimes encountered situations in which people were quite evidently unwilling to tell me things. At other times, I am sure that people steered me around topics that they sought to avoid without my noticing. Maneuvers of this sort, whether artful or awkward, are a ubiquitous feature of social interaction.[12] When I did encounter unretellable "secrets," I endeavored to treat their existence not only as a constraint on reporting data but also—like the conversation with Jay—as a form of data relevant to my research. Most of the time, my interlocutors seemed eager to describe their work and to share their views about the emerging world of genomics. I greatly appreciate their talking with me and telling me as much (or as little) as they did.

This account of the conversation with Jay is displayed using the conventions for preparing and presenting excerpts of edited field notes employed throughout the book. Such accounts appear in a sans serif font, with the names of individuals, universities, or companies replaced with pseudonyms (e.g., KATZ or UNIDENTIFIED INSTITUTE) or generic labels (e.g., FAMOUS GENETICIST) displayed in SMALL CAPS. (The small caps are used in only the first appearance of the pseudonym in each chapter.) In the case of text presented as quotations in excerpts from field notes ("Well, there aren't that many ideas out there"), I did my best to capture the exact language used. The original, unedited field notes were dictated or typed after the events that they depict, so there is clearly some room for slippage in the precise wording, even given my best efforts. The act of dictation was supported by whatever notes could be taken during or shortly after events, given the constraints of decorum and convenience. The process of creating accounts of the field notes presented in this book thus involved successive reconstructions, with an effort to preserve fidelity of meaning and language.

Working backward in time, these accounts consist of (6) *edited versions* of (5) *dictated or written field notes* based on (4) *memories*, often supported with (3) *written notes taken shortly after* and/or (2) *written notes taken contemporaneously* with (1) *events as I perceived them*. The goal of the editing was to convey the meaning of the original notes rather than to preserve their exact form. Editing (step 6) aimed to eliminate redundancy, clarify language, correct grammar, delete unnecessary details, smooth the narrative by reordering statements, and, when necessary, protect confidentiality by omitting identifying information. Ideas or statements not present in the original excerpt were sometimes added to supply context; for example, in the account about Jay, the phrase "people like him whose careers were not secure" was added even though I did not record it in the original notes because this fact was part of the known context within which the conversation took place. The insecurity of Jay's career at that time is evident elsewhere in my data.

Some of the events described in this book took place during the formal sessions of conferences with many people in attendance. The Cold Spring Harbor "Genome Mapping and Sequencing" meetings, the most important annual scientific conference of the HGP community, were particularly useful field research sites.[13] The Grace Auditorium at the Cold Spring Harbor Laboratory seats several hundred, and it was often jam-packed during these meetings; in some years, additional participants watched from other rooms by means of closed-circuit TV. In my field notes, I sometimes described statements or exchanges among the participants in open sessions of large meetings. These settings raise questions about identifying people because many people witnessed these exchanges, and some of them may clearly recollect who said what. Moreover, the organizers of some of these meetings recorded them. There is also a strong argument that gatherings bigger than some threshold are inherently "public" events, so rules of confidentiality do not apply. This issue came up quite pointedly at one of the very first meetings on molecular genetics that I attended. The meeting, held at a conference center outside a large American city, occasioned a dispute about whether it was legitimate to ask a journalist not to announce a particular scientific finding in his newspaper.

During a break, Dr. Montague, a scientist who was scheduled to give a talk later in the day, approached a table where Mr. Harold, a science journalist from a leading American newspaper, and several other people, myself included, were seated. Montague asked Harold to treat his talk as off the record. The reason: in his talk, he planned to briefly discuss an important scientific discovery that would be published and publicly announced in just a few days. A major disease gene had been found. This discovery, he said, is so significant that it would appear on the front page of the *New York Times* the day after it is announced. Montague wanted to present this finding to the conference without worrying that his comments would lead to a newspaper article. Harold objected, arguing that it would be inappropriate for him to promise to refrain from writing about something he learned at a meeting that he had been invited to attend as a reporter. Montague explained that although he was connected to the group that had made the discovery, he stressed that it was not "my work." He merely wanted to show a slide or two because he thought the discovery was relevant and important for the theme of the conference. Couldn't this one thing be treated as confidential for a few days? Harold replied that there is no such thing as a private meeting of 100 people. Montague said he was worried that a premature newspaper article could interfere with proper scientific publication. The *New England Journal of Medicine*, he noted, had a policy that it would not publish a scientific paper if its results had already been reported in the news media.[14] Harold considered the matter one of journalistic principle and would not bend. They argued for several minutes but could not resolve the impasse. Later, when Montague gave his talk, he announced that an extremely important gene had been found but that he could not name the gene or present the unpublished results, and he flashed through several slides at a pace clearly intended to be tantalizing yet completely unreadable.[15]

Harold makes a defensible moral and epistemic claim that a meeting of 100 people cannot be considered "private"—at least insofar as journalism is concerned. However, I decided to use pseudonyms even in instances where many people who were present might be able to recall who said what to an assembled group. Audience members also are never identified by name.

Documents

The world of genomics is full of writing and texts, and a variety of written documents were the third major source of data for this study. How I handled information from these sources and the extent to which I identify them in this book depends largely on their publication status. When drawing on journal articles, books, and other published materials, I provide full citations, crediting authors in the usual manner. Similar practices apply to text from newsletters, press releases, briefing books, packets passed out at meetings, and other documents produced for distribution. At times, I use information from unpublished personal letters, memoranda, and certain "internal" documents while maintaining confidentiality about their source. Such documents either serve as unreferenced background material or, if quoted or paraphrased, are treated like interview excerpts, with general information about the source provided in ways that do not compromise individual identities.

* * *

This discussion of my practices for gathering and presenting information serves to provide readers with a sense of how this study was conducted and to describe, in a bow to reflexivity, how my own knowledge-control practices are implicated in the gathering and reporting of the data presented here. The appropriateness of the word *implicated* was nicely underlined by an episode that took place at the end of a research trip to France. After attending a scientific meeting, giving a short talk about my work to a group of my interlocutors, visiting a genome laboratory, and conducting several interviews, I celebrated a successful week with dinner at a fine Parisian restaurant. Toward the end of the evening, the chef, clad in his traditional cap and a stained white apron, made a tour of the dining room, stopping at each table to banter with his guests. I was the only person seated alone, and after inquiring about how I had enjoyed the meal, he asked me where I was from.

"New York."

"So what is your business? What are you selling to the poor, poor French?"

"Nothing. I'm merely gathering information."

"A *spy*?"

"Of sorts."

We laughed, as did the couple at the next table whom I had noticed watching us. The chef recommended a dessert wine. Once he was gone, I found myself reflecting on the similarities and differences among field research, salesmanship, and espionage.

Notes

Chapter 1

1. Gilbert 1987, 26. See also Kevles and Hood 1992, especially the essays by Cantor, Caskey, Gilbert, Hood, and Watson.

2. Clinton 2000a.

3. I briefly discuss research methods later in this introduction and more fully in the appendix. Interviews and observations are supplemented by analysis of published and unpublished documents.

4. Kuhn 1962, 6–7.

5. Hacking 1999.

6. Kuhn 1962, 7–8.

7. Ibid., 111.

8. Fleck does not use the concept of a scientific revolution, but he leaves no doubt that he considers the focus of his inquiry (the Wassermann test for syphilis) to be an "epoch-making achievement" ([1935] 1979, 14).

9. Fleck argues that a thought collective develops whenever two or more persons are exchanging thoughts and notes that a stimulating conversation between two people may produce thoughts that neither of them would have had on their own or in different company (ibid., 44).

10. Ibid., 107, 111–113.

11. Löwy 1988, 147.

12. In an important footnote, Fleck expresses the view that changes in the broader social order are related to changes in scientific thought. He writes that great transformations in scientific thought style "often occur during periods of general social confusion. Such 'periods of unrest' reveal the rivalry between opinions, differences

between points of view, contradictions. ... A new thought style arises from such a situation. Compare the significance of the Early Renaissance or of the period following the Great World War" ([1935] 1979, 177 n. 4).

13. Latour argues that there is no better example of a revolution that has transformed society than the one attributed (at least in France) to Louis Pasteur (1988, 8).

14. Latour 1988; see also Latour 1987, 1993.

15. Jasanoff 2004a, 22–23.

16. Jasanoff 2004b; see also Shapin and Shaffer 1985 and Jasanoff 2005.

17. In their influential book on the controversy between Robert Boyle and Thomas Hobbes, historians of science Steven Shapin and Simon Schaffer delineate three ways in which the making of knowledge and the making of political order occupy the same domain: first, scientists create and maintain a polity within which they conduct their work; second, the products of that work "become an element in the political activity of the state"; and third, there is a "conditional relationship between the nature of the polity occupied by scientific intellectuals and the nature of the wider polity" (1985, 332).

18. Jasanoff 2004, 2005; see also Jasanoff 2003, 2012.

19. Studies of laboratories show the diversity of entities involved in knowledge production. See, for example, Latour and Wolgar 1979; Knorr Cetina 1981; H. Collins 1985; and Latour 1987. See also Hilgartner and Brandt-Rauf 1994; Rheinberger 1997.

20. These spaces may include what Andrew Barry (2006) calls "technological zones."

21. Gieryn 1983, 1999a.

22. Jasanoff 1987.

23. Gieryn 1999a; see also Andrew Abbott 1988.

24. Hohfeld 1917.

25. See, for example, Krasner 1982 for a widely cited definition of the term *regime*. See Foucault 1975 on regimes of truth. Regimes vary in scale, strength, and stability but must exhibit sufficient temporal persistence to provide a framework for guiding the action that they regulate for a time. It makes no sense to refer to a regime imposing rules on a game if the regime does not last through at least one "cycle of play." Hilgartner 1995 and Kling, Spector, and Fortuna 2004 use the concept of regime to discuss scientific communication.

26. The paradoxes of classification regimes are many. For example, Michael Aaron Dennis (1994) discusses the importance to classification of designating some items

as *unclassified*. Peter Galison (2004) describes the paradoxes stemming from efforts to control both isolated tidbits of knowledge and entire domains. See also Balmer 2006. On intellectual property, see, for example, Boyle 1996; Lessig 2004; Jasanoff 2005, 203–24; and Winickoff 2015. See also Proctor and Schiebinger 2008.

27. See, for example, Kelty 2008 on free and open-source software; Lessig 2004 on Creative Commons; and Hilgartner 2013 on BioBricks. On scientific authorship, see Biagioli and Galison 2003. See also Foucault 1984.

28. Jasanoff 1990; Hilgartner 2000.

29. See Kohler 1994; see also Daston 1995.

30. On collaboration, see Shrum, Genuth, and Chompalov 2007.

31. On "guided doings," see Goffman 1974.

32. The dual use of the construct *agent*, on the one hand, to frame generic arrangements and, on the other hand, to provide a frame to interpret specific situations also applies to the constructs *objects*, *spaces*, and *control relationships*.

33. Hohfeld 1917. Although Hohfeld's subject matter—legally recognized relationships—is not coterminous with the quasi-legal relationships found in many less-formal knowledge-control regimes, his insights about correlatives and the value of precision remain useful in the lawlike contexts examined here.

34. To find Hohfeld's basic insights helpful, we need not accede to his strong claim that his typology adequately captures all common legal relationships.

35. For an account of how AIDS activists contested control in human experimentation, see Epstein 1996.

36. Carruthers and Ariovich 2004.

37. See, for example, Ostrom 1990.

38. Kelty 2008; Stallman 2010. These forms can be conceptualized as common-pool resources; see Hess and Ostrom 2003. Jorge Contreras (2011) uses Ostrom's approach to describe efforts to create "common-pool resources" for genome research.

39. Callon 1994.

40. See, for example, Hilgartner 2004; Foray 2006; and Harvey and McMeekin 2007.

41. See Gusfield 1980 on public problems; Andrew Abbott 1988 on professions; Knorr Cetina 1999 on molecular biology; Jasanoff 1995 on decision making; and Boyle 1996 on domestic spheres of privacy.

42. Compare the work by Mark Harvey and Andrew McMeekin (2007), who offer an interesting analysis of the production of "public" and "private" "economies of knowledge" in bioinformatics from the perspective of economic sociology.

43. To be sure, knowledge-control regimes play a role in a variety of "economies," including various forms of commodity exchange, gift economies, credibility economies (Latour and Woolgar 1979); capitalist relations (Sunder Rajan 2006; Harvey and McMeekin 2007); reputational economies; promissory economies (Fortun 2008); economies of hope (Sunder Rajan 2006); moral economies (Kohler 1994; Daston 1995); and so forth.

44. Goffman 1974, 25. Related ideas about frames developed independently in sociology (ibid.), linguistics and cognitive science (Lakoff and Johnson 1980), and computer science (Minsky [1974] 1975). For a catalog of some of the frames embedded in language, see FrameNet n.d.

45. Ibid., 22.

46. Governing frames can emerge through a wide variety of "official" and ad hoc processes, including formal governmental action, informal negotiations, and even revolutionary acts. They may be produced by decree or emerge over time from subtle changes in widely distributed practices. Governing frames themselves are the constructed result of actions and practices, and they are produced, made manifest, and are transmitted in many ways. For example, people circulate idealized versions of them, teaching them, learning them, and building their identities and practices in ways that utilize them. Governing frames offer interpretive resources that actors can invoke, deploy, stretch, bend, and even break. The ready availability of the templates that these frames provide encourages actors to draw on them. Moreover, if well established in a particular historical and social context, a governing frame often has considerable obduracy, and actors may experience it as a constraint that imposes high costs on those who ignore or violate its strictures.

47. For an account of some of the contingencies involved in the election of Bush, see Lynch, Hilgartner, and Berkowitz 2005.

48. The meaning of the US Constitution, for example, has changed significantly over time, sometimes through constitutional amendments and sometimes through Supreme Court decisions, not to mention through war, especially the Civil War.

49. Entities are often reformatted immediately prior to transfer; for example, manuscripts are edited to conform to journal length constraints, and biomaterials are packaged and labeled. A standardized form may be executed to mark the moment of transfer, documenting the transformed status of the entity, assigning it a new identifier (such as an accession number), and registering it into the receiving regime's tracking system.

50. Smith, Sanders, Kaiser, et al. 1986; Smith, Hood, Hunkapillar, et al. 1992. See Cambrosio and Keating 1995 for an example of the transformation of an entity—hybridomas—from fact to technique.

51. Indeed, even within a single regime, the "same" entity, such as a human individual, often occupies different positions in different situations. The editor of a scientific journal is without a doubt also the author of published papers. This point relates to long-standing issues in science and technology studies about multiplicity, enactment, and ontology, some of which have been glossed as the "ontological turn" (see Mol 2002; Woolgar and Lezaun 2013).

52. See Jasanoff 2013 on problems of "epistemic subsidiarity," which arise when multiple levels of governance come into contact. See also Winickoff 2015.

53. The theoretical approach advocated here does not fully adopt the metaphysics of Latour's actor–network theory. In contrast to Latour, I reserve a special place for human agency but fully recognize the constructed character of the boundaries and agency of individuals, groups, and organizations. Temporarily bracketing the problematic aspects of human agency is a self-conscious "sociological" strategy for making the specific problems addressed here more tractable while remaining open to the possibility that analyzing some issues may require removing the brackets. See Lynch 1993, 107–113, for a discussion of sociological readings and misreadings of actor–network theory.

54. For the Mertonian sociology, see Merton 1973. See Mulkay 1976 and Shapin 2008 for critiques of the Mertonian perspective.

55. In this respect, the approach taken here is congenial with some flavors of the new institutionalism, with their emphasis on cognition, scripts, routines, and discourses; see Powell and DiMaggio 1991.

56. The items in this list sometimes overlap and sometimes can be combined in a single act.

57. Allowing a certain amount of "outlawed" activity is arguably socially beneficial (see Peñalver and Katyal 2010), and, in this view, making the regulatory capacity of knowledge-control regimes too effective has its downsides. See Lessig 2001 on the dangers of the erosion of the category of fair use in copyright.

58. One of the themes of Goffman's book *Frame Analysis* (1974) is the centrality in everyday social interaction of fabrications designed to foster wrong impressions about the nature of the situation.

59. Stallman 2010.

60. Classic works include Latour and Woolgar 1979; Knorr Cetina 1981; Lynch 1985; and Traweek 1988. See also Knorr Cetina 1999; Kleinman 2003; Doing 2009; and Nelson 2013.

61. This series of annual meetings was later retitled "Genome Sequencing and Biology" to reflect a shift in the HGP's focus to sequencing.

62. I also attended the Human Genome I and Human Genome II conferences, two chromosome-specific scientific workshops, a Chromosome Coordinating meeting, several advisory committee or program policy meetings, several National Research Council workshops related to genomics, and a number of other policy-oriented conferences and workshops. In addition, I served on several NIH grant review panels for its genome program and on a committee that advised the NCHGR and the US DOE on its program on the ethical, legal, and social implications of the genome project. The confidential discussions and interviews connected with these committee activities are not quoted or paraphrased here.

63. Cook-Deegan 1994a. For scientists' accounts, see Sulston and Ferry 2002 and Venter 2007; on science writers, see K. Davis 2001, Zweiger 2001, Shreeve 2004, and McElheny 2010. See also Glasner and Rothman 2004.

Chapter 2

1. Steven Shapin begins a book with the startling statement that "there was no such thing as the Scientific Revolution, and this is book about it" (1996, 1).

2. Tilly 1978, 189.

3. Tilly, Tarrow, and McAdam 2006.

4. Even the most neutral descriptions of revolutions tend to be examples of what Austin calls performative speech acts (Austin [1962] 1975; see also Skinner 1969).

5. On the hybrid domain, see Latour 1993.

6. My account does not rest on using the "genomics revolution" as an analysts' category, but it is crucial to recognize its importance as an actor's category.

7. Some of my interlocutors often used the term *audacious*, inflected with a positive valence.

8. See, for example, Turner 2006 on digital utopianism and McCray 2012 on space colonization and nanotechnology. For business literature celebrating "disruptive technology," see Christensen 1997 as well as Day and Schoemaker 2000. See Edgerton 2007 for a critique of an innovation-centric vision of technology.

9. Hilgartner 2015.

10. Jasanoff and Kim 2015.

11. Members of vanguards are typically drawn from such often overlapping categories as "promise champions" (van Lente and Rip 1998), entrepreneurs and startup companies (e.g., Sunder Rajan 2006; Fortun 2008), bioethical analysts (Hedgecoe and Martin 2003), and social movements or "recursive publics" (Kelty 2008), such as the promoters of free and open software (Coleman 2009). Policy entrepreneurs,

inside or outside of government, can also play central roles in sociotechnical vanguards, as the histories of such military-industrial complex projects as the H-bomb and the ARPANET (Advanced Research Projects Agency Network) illustrate. These vanguards aim to shape expectations, inspire hopes and fears, and bring specific futures into being. See, for example, Brown, Rappert, and Webster 2000; Fortun 2001, 2008; Hedgecoe 2004; Rabinow and Dan-Cohen 2005; and Tutton 2011.

12. Gieryn 1999a.

13. Hajer 2009.

14. Early proposers of a sequencing project included Robert Sinsheimer, president of the University of California, Santa Cruz; DOE scientists Charles DeLisi and David Galas; Renato Dulbecco, president of the Salk Institute; and, in the United Kingdom, Sydney Brenner of the MRC. The vanguard soon expanded, with Galas and DeLisi working to launch a DOE program and growing numbers of science policy discussions. James Watson, Walter Gilbert, and others were soon involved. See Cook-Deegan 1994a.

15. Finding a disease gene entails first identifying the region of the genome where it is located and then honing in on it via detailed mapping of that region. See, for example, Rommens, Iannuzzi, Kerem, et al. 1989.

16. W. Gilbert 1987, 29. See also W. Gilbert 1992.

17. Cook-Deegan 1994a offers the most comprehensive available account of the origins of the HGP.

18. NRC 1988. See also Office of Technology Assessment 1988. On model organisms, see de Chadarevian 2002, 2004; Ankeny and Leonelli 2011; Nelson 2013; and Rader 2004.

19. Four people were centrally involved in the research leading to this discovery, although Watson and Francis Crick received the bulk of the credit. The Nobel Prize was awarded to three of them (Crick, Watson, and Wilkins) in 1962. Rosalind Franklin, who produced images using x-ray crystallography that were crucial to discerning DNA's molecular structure, died of cancer in 1958. The Nobel Prize is not awarded posthumously.

20. See Joly and Mangematin 1996 and Hilgartner 1998 on the European yeast program. See Rabinow 1999, Kaufmann 2004, and Rabeharisoa and Callon 2004 on Généthon. See Balmer 1996a, 1996b, 1998 on the UK HGMP. See also Jordan 1993 for a view of several programs in the US, Europe, and Japan.

21. The International Human Genome Sequencing Consortium consisted of a number of major centers for genome research that participated in data generation and bioinformatics. The laboratories were: Whitehead Institute/MIT Center for

Genome Research, Cambridge, Mass.; Wellcome Trust Sanger Institute, Wellcome Trust Genome Campus, Hinxton, Cambridgeshire.; Washington University School of Medicine Genome Sequencing Center, St. Louis; DOE Joint Genome Institute, Walnut Creek, Calif.; Baylor College of Medicine Human Genome Sequencing Center, Department of Molecular and Human Genetics, Houston; RIKEN Genomic Sciences Center, Yokohama; Genoscope and CNRS UMR-8030, Evry, France; Genome Therapeutics Corporation Sequencing Center, Waltham, Mass.; Department of Genome Analysis, Institute of Molecular Biotechnology, Jena, Germany; Beijing Genomics Institute/Human Genome Center, Institute of Genetics, Chinese Academy of Sciences; Multimegabase Sequencing Center, Institute for Systems Biology, Seattle; Genome Technology Center, Stanford University, Stanford, Calif.; Human Genome Center and Department of Genetics, Stanford University School of Medicine; University of Washington Genome Center, Seattle; Department of Molecular Biology, Keio University School of Medicine, Tokyo; University of Texas Southwestern Medical Center at Dallas; University of Oklahoma Advanced Center for Genome Technology, Department of Chemistry and Biochemistry, Norman; Max Planck Institute for Molecular Genetics, Berlin; Cold Spring Harbor Laboratory, Lita Annenberg Hazen Genome Center, Cold Spring Harbor, NY; and GBF—German Research Centre for Biotechnology, Braunschweig, Germany.

22. To further complicate matters, a group of pharmaceutical companies established the SNP Consortium, which provided additional funding to the "publicly funded" HGP during the competition with Celera (by supporting a project to find single-nucleotide polymorphisms and make them freely available).

23. W. Gilbert 1991, 99; see also Hood 1992.

24. Field notes, 1990.

25. Bruno Strasser (2011) argues that the sequence databases and their centrality suggest that we should rethink the standard historiographic narrative of the triumph of experimental science over natural history in biology because they demonstrate the importance of collections and collecting practices. Strasser also emphasizes the formation of networks among scientists and moral economies associated with collections. For a comparison of collecting practices in natural history with those of genome mapping, see Stemerding and Hilgartner 1998.

26. See Hine 2006 on databases as instruments and Rheinberger 1997 on epistemic things.

27. See Leonelli 2011, 2012; Stevens 2013. James Fickett, a mathematician who was centrally involved in setting up GenBank, argued in the late 1980s that databases had become a communication network "analogous in function to scientific meetings, journals, or the international telephone system" (1989, 295). See also Hilgartner 1995.

28. Stevens 2013.

29. Roberts 1987. The potential conflict of interest led to Gilbert's resignation from the NRC Committee that wrote *Mapping and Sequencing the Human Genome.*

30. See Fortun 2008 for an analysis of the promises of genomics.

31. McKusick and Ruddle 1987, 1.

32. Gaudillière and Rheinberger 2004.

33. See, for example, Nukaga and Cambrosio 1997; Terry 2003; and Rabeharisoa and Callon 2004.

34. Fleischmann, Adams, White, et al. 1995. According to Michael Cinkosky, James Fickett, Paul Gilna, and Christian Burks, "About 2850 organisms (including viruses) are represented in GenBank. The only completely sequenced genomes are from viruses and cell organelles (mitochondria and chloroplasts), ranging in size from a few hundred base pairs for certain plant viruses to more than 200 kilobase pairs for the cytomegalovirus" (1992, 272).

35. For example, Davis and Colleagues 1990.

36. Human cells contain 46 chromosomes: two copies of chromosomes 1 through 22 plus either two X chromosomes (in females) or an X and a Y chromosome (in males). The 46 chromosomes contain 6 billion base pairs of DNA, but a sequence of the 24 unique chromosomes is only 3 billion.

37. NRC 1988, 56.

38. Ibid., 62, 65.

39. Ibid., 6.

40. Radiation hybrid maps were another kind; see Cox, Burmeister, Price, et al. 1990.

41. NRC 1988, 42–43.

42. These maps took advantage of "restriction enzymes," which cut through a DNA molecule wherever it encounters a specific short sequence, such as GCGGCCGC. Cutting a DNA molecule with a restriction enzyme and measuring the distance between the cuts creates what is known as a "restriction map."

43. Smith, Econome, Schutt, et al. 1987; see also Kohara 1987.

44. NRC 1988, 2–3.

45. The same obsessions remain prominent today, although data production is many orders of magnitude faster.

46. Fortun 1998.

47. NRC 1988, 2–3.

48. On laboratory automation, see Keating, Limoges, and Cambrosio 1999.

49. Field notes, 1989.

50. NRC 1988, 76.

51. Baltimore 1987, 49. See also Kevles 1997.

52. NRC 1988, 101.

53. See, for example, Holtzman 1989; Nelkin and Tancredi 1989.

54. See, for example, Duster 1990; Lewontin 1991; Beckwith 2002.

55. Watson 1990, 46. The DOE followed suit shortly thereafter. The percentage later grew to 5 percent.

56. Juengst 1994, 121.

57. Collingridge 1981. See also Joly 2015.

58. For analyses of the politics of programs like ELSI, see Juengst 1996 and Hilgartner, Prainsack, and Hurlbut in press. See also Rabinow and Bennett 2012.

59. Juengst 1996, 64, 66.

60. Hilgartner, Prainsack, and Hurlbut in press.

61. The main exception was ELSI research on the use of human subjects in genetic research. In addition, Eric Juengst (1996) describes several examples of ELSI-funded studies that looked critically at the HGP. The separation between the science and its implications was sharpened after Francis Collins, director of the NCHGR, eliminated a "working group" of outside advisers and consolidated control over the program in the NIH staff (Hilgartner, Prainsack, and Hurlbut in press).

62. Knorr Cetina 1999. The molecular biology laboratories that Knorr Cetina examined were in an institute in Germany that drew scientists from across Europe, but her findings accord well with what I observed in US laboratories at the outset of the HGP. Her book was published in 1999 based on data collected mainly in the 1980s and early 1990s; it was the first ethnographic study of scientific laboratories to compare two fields (molecular biology and high-energy physics).

63. On craft work in molecular biology, see Fujimura 1996 and Knorr Cetina 1999; see also Clarke and Fujimura 1992.

64. See Jordan and Lynch 1992 for a discussion of the need for local adaptation of the "standard" technique plasmid prep.

65. Field notes, 1988.

66. Knorr Cetina 1999.

67. The custody thus maintained was much looser than what is required to construct the "chain of custody" required for legal use of DNA forensic materials; see Lynch, Cole, McNally, et al. 2008.

68. Some inscriptions, such as photographs, were pasted into notebooks, but those that did not fit were tagged with identifiers intended to link them to corresponding notebook entries.

69. In the 1980s, several cases of scientific fraud and misconduct were accompanied by shoddy record keeping; see Kevles 1998.

70. Bertrand Jordan describes the repetitive work involved in mapping with similar techniques as "mindnumbing [*sic*]" (1993, 22).

71. Knorr Cetina 1999.

72. Church and Kieffer-Higgins 1988.

73. The data by themselves were not really enough to complete a definitive evaluation. The reason: evaluating a sequencing strategy in this way suffers from what Harry Collins (1985) calls the "experimenter's regress." If a laboratory fails to make a sequencing strategy work, it may be because the strategy has problems or because the laboratory has implemented it poorly.

74. See García-Sancho 2012 for an informative account of the development and commercialization of the ABI instrument.

75. They carefully poured gels, working to avoid the formation of bubbles, which would disrupt the migration of DNA within the gel. Then they had to load samples into the gels. With traditional gel-based sequencing, each and every base in the sequence had to be "called" (that is, identified as an A, C, G, or T) and entered into a computer after the gel was "run." The sequencing machines introduced at the end of the 1980s automated this base-calling and data-entry step, which previously was done using human hands and eyes. Human beings still did everything else, though.

76. For example, runs of the bases C and G (e.g., CCCGGCG) tended to produce so-called compression errors because gels could not neatly separate strings of Cs and Gs into individual bases.

77. Field notes 1990, 1991, 1992 and interview 1992.

78. Ibid.

79. Interview 2000.

80. Ibid.

81. Several observers have written useful accounts of the changes in sequencing technology that enabled genome scientists to build these new factorylike

sequencing facilities; see Hutchinson 2007; Rowen 2007; García-Sancho 2012; and Stevens 2013. An extended look at these changes is beyond the scope of this book, but the main developments include system integration, especially the integration of sequencing technology with computer systems; the introduction of improved sequencing machines; the computerization of key steps, such as better base calling and sequence assembly by software; the introduction of robotics to perform slow mechanical steps; the development of sequencing strategies; the simplification of laboratory protocols; the routinization of work and its shift to technicians; the scaling up of the size of laboratories; the development of improved cloning vectors, such as BACs; the engineering of better enzymes for use in sequencing; and, not least, the ongoing increase in the power of information and computer technology. Since 2000, a new generation of sequencing technologies that do not rely on gel electrophoresis has been introduced, greatly increasing throughput.

Chapter 3

1. None of the laboratories discussed in this chapter focused on sequencing, which was not the HGP's main short-term goal at this time.

2. Knorr Cetina 1999.

3. How laboratory heads use their privileges vary, creating differences in the treatment of subordinates and colleagues. Molecular biologists typically—and tellingly—explain differences in terms of laboratory leaders' "personalities."

4. It is hard to generalize about how researchers assess strategic value, but several qualities seem often to figure in their evaluations. First, holdings associated with central laboratory goals tend to be considered more valuable than those pertaining to "side projects." Second, entities that researchers viewed as foci of research tended to have higher strategic value than "infrastructure," as defined in Star and Ruhleder 1996 and Star 1999. Third, unique or rare entities might convey a competitive advantage and tended to be considered more valuable (Fujimura 1987; Hilgartner and Brandt-Rauf 1994). However, bringing new entities into a laboratory entailed work and risk, so tried-and-true holdings had value because their limitations were understood. Incorporating new techniques, machines, and materials into a laboratory entailed work and risk, so laboratories tested them to see how they work "in our hands." Kathleen Jordan and Michael Lynch (1992) point out that even the most standardized techniques must be adapted to local sites (see also Lynch, Cole, McNally, et al. 2008). Laboratories introduced novel entities cautiously and exchanged information on what "worked well." Finally, *when* assessing the strategic value of holdings, researchers often considered the degree of certainty about them. Certainty, of course, could affect strategic value of a knowledge object either negatively or positively, raising it if actors were sure that they could place confidence in it, reducing it if the object were known to be unreliable. Claims about the level of

uncertainty are sometimes best understood as a form of strategic action used to persuade audiences. See, for example, Pinch 1981, 1986; Campbell 1985; and Murphy 2006. See also the analysis of the experimenter's regress in H. Collins 1985 and the discussion of credibility in Latour and Woolgar 1979.

5. Kopytoff 1986.

6. Races among parties intent on a single well-defined goal produce tightly "focused competition" that closely resembles a zero-sum game. This situation differs greatly from "diffuse competition," as in the unfocused struggle of each against all for research funds.

7. On region behavior in interaction, see Goffman 1959.

8. Galison and Thompson 1999; Gieryn 1999b. Some university life sciences buildings are now equipped with special areas set apart from the bulk of the laboratory space and sealed off for extra security, where faculty members can develop ideas that they imagine commercializing free from prying eyes and leading questions.

9. Shapin 1988.

10. Within the dominion of the laboratory, access to knowledge objects is further compartmentalized. The laboratory head may not share all of his or her ideas or plans with other personnel, and access to certain data, materials, and texts may also be controlled.

11. Some of these transformations, such as wrapping electronic data in metadata or editing texts to meet length constraints, are intended to enable circulation; see Howlett and Morgan 2011.

12. This practice, although perhaps most commonly associated with bureaucratic and legal procedure, is also commonplace in social research, including this book, as discussed in the appendix.

13. Hilgartner 2000, 53.

14. Approaches that treat knowledge objects as essentially interchangeable "resource units," such as those building on the work of Elinor Ostrom on common-pool resources, miss the process through which a perimeter can be built into knowledge objects. For an example, see Hess and Ostrom 2003.

15. These three contexts serve as a way to organize my account; they are not wholly distinct forms of interaction.

16. Survey research (e.g., Campbell, Clarridge, Gokhale, et al. 2002) provides broad, if superficial, data about the prevalence of some of the practices described here.

17. Informal speech refers to "ordinary" conversation rather than to such specialized forms of talk as the lecture; see Goffman 1981.

18. Laboratories, of course, regulated such visits. One genome center, concerned about the burdens of hosting short-term visitors, enacted a policy of allowing visits of only two lengths: either one day or one year (to do a collaborative project).

19. On information control in interaction, the classic source is Goffman 1959.

20. Field notes, 1989.

21. Knorr Cetina (1999) found joking to be common in molecular biology, where it seemed to ease tensions around hierarchy and other minor conflicts.

22. In most companies, Mario and Rich would probably have had to sign nondisclosure agreements before leaving, and Rich would not be allowed to simply walk back into the laboratory after accepting employment with a competitor, whether to say good morning or anything else. But such moves would have seemed completely out of place in this academic environment. James Evans (2010) argues that although both academic and industry scientists keep secrets, industry is better at doing so.

23. For some examples and a discussion, see the appendix.

24. Field notes, 1991.

25. For examples of similar maneuvering in the context of animal research, see Holmberg and Ideland 2012.

26. Galison 2004.

27. Michael Aaron Dennis (1999) makes this point regarding a military research context.

28. The author enjoys much discretion to make threshold judgments about impracticality, and there is no third party to review such judgments.

29. Over time, more formal Material Transfer Agreements grew increasingly common, and many organization's policies mandated them, especially when industry was involved. See Mirowski 2011.

30. Vollrath, Foote, Hilton, et al. 1992, 59.

31. BETH, interview, 1989. In displaying quotations from interviews, ellipses indicate elided material. Brackets are occasionally used to insert a clarification. Pauses in speech are not documented.

32. Ibid.

33. One laboratory head told me he had "tech working half-time" to supply requesters with a particularly useful clone until finally he said, "This is ridiculous," and arranged for American Type Culture Collection to send it out for $100 (interview, 1991).

34. Erin, interview, 1989.

35. This account is based on interviews and conversations with the laboratory head, the postdoc who ran the project, and technicians in 1992.

36. In fact, the laboratory did distribute the map with what genome researchers considered minimal delay.

37. Jon, interview, 1992; field notes and interviews with other Marigold personnel, 1992.

38. The 365-seat auditorium at Cold Spring Harbor was often packed. On some occasions, closed-circuit TV piped the event to overflow rooms. In poster sessions, researchers displayed research in a visual format mounted on a wall or bulletin board. Audience members would wander through rooms where presenters waited beside their posters for inquiries and conversations. Poster presentations, though less important, were far from insignificant. Limited plenary slots dictated that many attendees and much high-quality work appeared only in posters.

39. Soraya de Chadarevian makes a similar point (2002, 287–299). The social structure and moral economy of the *C. elegans* community of the late twentieth century resembled in several crucial aspects (size, interdependence, and personal connections) the *Drosophila* community of the early twentieth century Robert Kohler studied in 1994.

40. In the early 1990s, I attended two SCWs as well a related policy-oriented meeting intended to coordinate this kind of research. The larger SCW had roughly 90 attendees, the smaller about two dozen.

41. Both SCWs that I attended included sessions that explicitly addressed the importance of data sharing, which underlines the problematic nature of data access in the field.

42. Field notes, 1992.

43. People moved in and out of the room, so the count varied. About 10 of the participants were women.

44. Field notes, 1992.

45. Ibid.

46. Field notes, 1992.

47. HUGO 1992.

48. On occasion, the term *collaboration* referred to joint projects within the same laboratory.

49. Grant proposals also frequently involved collaborations, which were often set up as a subcontract between the principal investigator's university and that of the collaborator.

50. The collaboration between Robert Waterston and John Sulston is a good example; see Sulston and Ferry 2002 as well as de Chadarevian 2004.

51. Wesley Shrum, Joel Genuth, and Ivan Chompalov (2007) provide a careful analysis of collaboration in such fields as particle physics, space sciences, and oceanography, areas of science in which collaborations are sometimes complex and bureaucratized.

52. This is consistent with the findings in Knorr Cetina 1999 and, in many respects, with those in Shrum, Genuth, and Chompalov 2007 in very different fields.

53. See Shrum, Genuth, and Chompalov 2007.

54. Patrick, Southern Genome Center director, 1992.

55. Jon, interview, 1992.

56. This is consistent with Steven Shapin's (2008) argument for the importance of face-to-face relations and personal familiarity in the supposedly impersonal modern world.

57. KATHLEEN, interview, 1991.

58. MIKE, interview, 1991.

59. See also Knorr-Cetina 1999 on the difficulty of collaboration in molecular biology.

60. Kathleen, interview, 1991.

61. See the example in Hilgartner 2012.

62. Paul Atkinson, Claire Batchelor, and Evelyn Parsons (1998) describe this dynamic in their study of the hunt for the myotonic dystrophy gene.

Chapter 4

1. The head of the Genome Analysis Laboratory, Hans Lehrach, began this work at the EMBL in Heidelberg before joining the ICRF.

2. The RLS and the US regime were not the only ones that emerged at this time. Other significant regimes include those governing Généthon in France (Rabinow 1999; Rabeharisoa and Callon 2004; Kaufmann 2004), the European Community effort to sequence yeast (Joly and Mangematin 1998), and the HGMP and Resource Centre set up by the UK MRC (Balmer 1996a, 1998). All of these regimes differed from both the US regime and the RLS.

3. OLLIE, interview, 1992.

4. On the process through which the US program won this commitment, see Cook-Deegan 1994a.

5. NRC 1988.

6. However, the producers of data expected also to be among the users, so individual scientists expected to play both parts.

7. Balmer 1996a reports that the HGMP also engaged in ongoing work to define what fell under the program's remit.

8. Official pronouncements to this effect were made both along the way and after the project was completed. See, for example, Collins and Galas 1993; Collins, Morgan, and Patrinos 2003.

9. For example, more than 80 percent of the people identified as contributors (e.g., because they served on a committee) to the report (NIH and DOE 1990) outlining the first set of five-year goals for the US genome program were men, as were 14 of the 15 members of the NRC Committee on Mapping and Sequencing the Human Genome (table 2.1) (NRC 1988). On advisory board members to the first genome companies, see table 5.1.

10. NIH and DOE 1990.

11. L. Roberts 1989a.

12. NCHGR 1989; Office of Human Genome Research 1989.

13. NCHGR 1989, 9.

14. Ibid., 11.

15. For an account of some of the mapping methods in the *C. elegans* (worm) community, see de Chadarevian 2004; see also de Chadarevian 2002.

16. Olson, Hood, Cantor, et al. 1989. In an accompanying news article, genome project leaders described the STS proposal as the "centerpiece" of the five-year plan (L. Roberts 1989b). In addition to the four named authors, Cassandra L. Smith also contributed to development of the STS concept (Cook-Deegan 1994a).

17. TOM, interview, 1991.

18. Even if people intended to share clones quickly, doing so raised the logistical problems associated with packing and shipping biomaterials. With increasing scale, the logistical complexities were expected to grow much worse, perhaps requiring the establishment of a repository capable of distributing hundreds of thousands of clones (NRC 1988, 81–82). The authors of the STS proposal were also worried about the long-term stability of clones, which, after all, are living cells known to spontaneously drop or to rearrange the DNA of interest.

19. To use an STS-based map to establish the location of a random DNA fragment on the genome, a laboratory needed only written inscriptions. An oligonucleotide synthesizer could translate the text of the STS landmarks into physical primers—that is, the corresponding DNA molecules. Laboratories could manufacture their own primers or buy them from companies that made them as a commercial service. Once converted into molecules, the primers could be loaded into a PCR machine, along with the sample to be tested, and run, yielding a positive or negative result in a few hours. The informatization of mapping landmarks also accounted for their ability to bridge across maps of different types made using different technologies, for all the maps described the same underlying DNA to which the sequence-tagged sites pointed.

20. Ollie, interview, 1996.

21. NIH and DOE 1990.

22. The highly successful French genome center, Généthon, which was supported with funds raised privately through a telethon, constituted itself along similar lines, distributing the resources that it created, such as genetic markers and maps, to any and all without encumbrances. See, for example, Généthon 1992.

23. A number of American mapping projects utilized the same libraries of YACs.

24. See the discussion of increasing efficiency in NRC 1988.

25. Field notes, 1992.

26. Tom, interview, 1991.

27. NCHGR 1992b. In contrast, the United Kingdom's HGMP lacked a metric for evaluation; a count of genes mapped, for example, was not used; see Balmer, Davidson, and Morris 1998.

28. Productivity in mapping was considered harder to conceptualize and measure than productivity in sequencing. The amount of sequence could be measured in base pairs completed, which genome researchers tended to regard as the "natural" unit of production (although, as we will see, this measure is not a straightforward metric for quantifying progress toward a "final," "completed" sequence).

29. See NCHGR 1992a, tab G, attachment 2.

30. Several scientists, especially those from the United Kingdom, expressed this view to me. The published comments by Peter Little, an advocate of the RLS who was unconvinced about the efficiency of STSs, in a commentary in *Nature* are illustrative: "STSs are *a* standard and not *the* standard" landmark (Little 1991, 21, emphasis in original).

31. Rabinow 1996 provides an account of PCR's development and incorporation into practice.

32. NIH and DOE 1990, 14.

33. See the discussion of the transmission of tacit knowledge in H. Collins 1985.

34. This extra step was not required in the case of some mapping strategies that were based on PCR. Scientists using other techniques to build maps, however, had to translate them into STSs.

35. EUROPEAN GENOME SCIENTIST, interview, 1992.

36. Eric Green, Rose Mohr, Jacquelyn Idol, and their colleagues published a paper in 1991 showing that STSs developed at Washington University worked at the Los Alamos genome center without any communication beyond conveying such publishable information as the sequences of primers.

37. NATHAN, interview, 1992.

38. This statement is from a version of the policy published in 1995; see Whitehead Institute/MIT Center for Genome Research 1995 (author's copy).

39. Ibid.

40. The data-release policy distinguished between uses of the data that would and would not require negotiations: "For projects aimed at the analysis of particular genes or subchromosomal regions, permission is hereby granted to use our data without the need for a formal collaboration, subject only to appropriate acknowledgement. For projects aimed at large-scale mapping of entire chromosomes or entire genomes, use of the data and markers should be on a collaborative basis" (ibid.). The latter kind of project, large-scale mapping, is the type of work that CGR itself performed.

41. The boundaries of peer review are sometimes contested, as Jasanoff 1987 shows.

42. Fujimura 1987.

43. Nathan, interview, 1992.

44. On laboratory automation, see Keating, Limoges, and Cambrosio 1999. See also Stevens 2013 on the sequencing factories that were eventually assembled.

45. Nathan, interview, 1992.

46. Ibid.

47. On these programs, see Balmer 1996a; Hilgartner 1998; Joly and Mangematin 1998; Rabinow 1999; Kaufmann 2004; and Rabeharisoa and Callon 2004.

48. Maier, Hoheisel, McCarthy, et al. 1992.

49. The RLS also explicitly declined to provide large numbers of clones produced using probes containing "repeats"—sequences that appear many times in a genome.

50. Zehetner 1991, 8, emphasis in original.

51. By perusing the database, a scientist could learn the tantalizing fact that E. M. VAN GALDER possessed clones using 21 different "confidential probes" and that confidentiality was set to expire on July 29, 1991 (extendable by request till January 21, 1992).

52. Zehetner 1991, 3.

53. The reference library approach immediately yielded DNA to study, which some saw as a distinct advantage over the STS approach (Jordan 1993, 127–128).

54. A number of US genome scientists I spoke with at the outset of the HGP expressed this view, which they view history as having borne out. Some advocates of the RLS remained unwilling to concede this point, arguing that the subsequent development of microarray technology and the lack of investment in improving Lehrach's methods raise doubts about the intrinsic superiority of PCR (interviews, 2001). Lehrach's work contributed to the development of "microarrays" that used hybridization to produce massive amounts of data in a single experiment. On the development of microarrays, see Lenoir and Giannella 2006.

55. LEW, interview, 1993.

56. See Zehetner 1991.

57. Lew, interview, 1993. US scientists were also impressed by Lehrach's automation of high-density arrays.

58. Interviews, 1992.

59. Zehetner 1991, 2.

60. Ibid., 2–3, emphasis in original.

61. Interview, 2001.

62. See B. Anderson 1991. Benedict Anderson focuses on the nation-state, but the concept of an imagined community is applicable to other kinds of collectivities.

Chapter 5

1. On the role of law, regulation, and ethics in constituting knowledge and order, see especially Jasanoff 2005.

2. This alternative approach was discussed as early as a Cold Spring Harbor symposium in 1986, where the idea of an HGP was debated; see Cook-Deegan 1994a, 315. See Kay 2000, Keller 2000, and Moss 2004 for explorations of the concept of the gene. See also Barnes and Dupré 2008.

3. Brenner 1990, 9. Estimates of the percentage of the genome coding for proteins included 2 percent (ibid.) and 5 percent (NRC 1988, 57).

4. NRC 1988, 57–58.

5. See Balmer 1996a, 1996b on the UK HGMP.

6. Davis and Colleagues 1990. Similar arguments were made at an NIH–DOE strategy meeting in Hunt Valley, Maryland, that I attended in 1993.

7. A project that the DOE funded included a Los Alamos initiative, some work on making better cDNA libraries, and some work by external investigators, including Venter. Researchers also began using cDNA sequences to develop STSs that pointed directly to genes (see, e.g., Wilcox, Khan, and Hopkins, et al. 1991).

8. NIH Ad Hoc Committee 1990, 2–3.

9. Fujimura 1996.

10. Venter 2007, 121.

11. Other databases were also used for such searches, including the EMBL sequence database and Protein Information Resource.

12. Venter 2007, 124. In his autobiography, Venter uses the term *discovering* (see also p. 122), which equates finding an EST with finding a gene. This equation was not advanced in the initial paper describing ESTs in 1991, which used the term *tagging* and said that ESTs had "applications in the discovery of new human genes" and that "the EST approach would provide a new resource ... for human gene discovery" (Adams, Kelley, Gocayne, et al. 1991, 1651, 1656). The term *finding* immediately came into common use, however.

13. Adams, Kelley, Gocayne, et al. 1991. This count of new genes was what was left after eliminating redundancies.

14. Ibid., 1651.

15. L. Roberts 1991, 1618.

16. L. Roberts 1991, 1619. Roberts quoted John Sulston as saying that Venter might get more like 8 to 9 percent. But see Sulston and Ferry 2002, 107, where Sulston says that the 8 to 9 percent number was a joke, impoliticly made to a reporter, and he meant merely to characterize 80 to 90 percent as optimistic.

17. Brannigan 1981; see also Cozzens 1989.

18. Field notes, 1991.

19. For a discussion of tools and jobs in science, see Clarke and Fujimura 1992.

20. For a retrospective example of Venter's responses to criticisms, see his autobiography (Venter 2007). Venter's supporters also made contemporaneous statements to me along the same lines.

21. On intellectual-property policy, see, for example, Boyle 1996; Lessig 2001; Jasanoff 2005; and Mirowski 2011.

22. See Winickoff 2015 for a discussion of the simultaneous reconfiguration of the public/private and nature/culture boundaries.

23. It is possible to argue that the EST is an "object" and the "EST patent" is merely an interpretation of that object that judges the EST to fit the criteria of patentability. While this is a reasonable view, the law is better understood as being capable of constituting entities, not just interpreting them. For example, forming a legal entity such as a corporation looks more like the addition of a new object to the world than like a mere interpretation. Similarly, EST patents can thus be profitably understood as new objects with different properties than ESTs. See Woolgar and Lezaun 2013 for a discussion of related issues.

24. An attorney from Genetech informed Adler about Venter's work and suggested it might be patentable (Cook-Deegan 1994a, 317; Venter 2007, 130).

25. Adler 1992. The idea that ESTs might be patentable was plausible given the expanding definition of patentability during the 1980s. On this expansion, see Boyle 1996 as well as Jaffe and Lerner 2004.

26. Cook-Deegan writes that "the decision to file the patent application was made by Venter's group, senior administrators of his institute (the National Institute of Neurological Disorders and Stroke), the Patent Policy Board at NIH, and the NIH Office of Technology Transfer, headed by Adler" (1994a, 317).

27. Cook-Deegan 1994a, 310–311. Cook-Deegan notes that Watson and NCHGR staffer Robert Strousberg had been previously informed.

28. See, for example, Eisenberg 1990 and Karjala 1992; see also Kieff 2003.

29. By February 1992, Venter was reportedly generating some 144 ESTs per day and predicting that the pace would only accelerate (Wheeler 1992b).

30. L. Roberts 1992, 912–913; Wheeler 1992a; see also Adams, Dubnick, Kerlavage, et al. 1992.

31. The ABI machines at that time had 24 lanes and could be run twice per day, with a theoretical throughput of 48 ESTs per day. A dozen such machines running five days a week could produce 2,880 ESTs per week. Even running at 25 percent capacity, such a system could generate some 37,000 ESTs per year.

32. A database search using an EST might allow inferences about putative functions.

33. Gillis 1992.

34. For example, during the summer of 1991 I happened to interview an HGP scientist who was leading a cDNA-sequencing project after he had learned of the patent application. He explained that he didn't have much information yet, but his immediate reaction was clear: patenting cDNA fragments was "crazy" and would cause "tremendous grief" to the genome project (PETE, a cDNA SEQUENCER, interview, July 1991).

35. During this period, a number of genome scientists I spoke with expressed dismay about what they regarded as the deficiencies of the legal system. The patent law seemed insensitive to the distinction between the routine and the creative, for example. Rights might be awarded in ways that made no "scientific" sense.

36. Pete, a cDNA sequencer, interview, July 1991.

37. Mathews 1991.

38. Quoted in Cook-Deegan 1994a, 311–312.

39. Quoted in L. Roberts 1992, 913.

40. Ibid.

41. Cook-Deegan 1994a provides a detailed account of Watson's resignation.

42. On the performative nature of claims about the future, see van Lente and Rip 1998; Brown, Rappert, and Webster 2000; Fortun 2008; and Gusterson 2008. See Austin [1962] 1975 on speech act theory.

43. Balancing the interests of past, present, and future inventors is a key aspect of intellectual-property policy; see, for example, Boyle 2003.

44. NATHAN, interview, 1992. HUGO later described "the scientific work involved in generating ESTs" as "straightforward" and based on mid-1980s technology. In contrast, using an EST to obtain a full-length cDNA could take from weeks to more than a year. "The task of identifying biological functions of a gene," HUGO argued, was "by far the most important step in terms of both its difficulty and its social benefit." "It would be ironic and unfortunate if the patent system were to reward the routine while discouraging the innovative" or "if a partial sequence publication or submission to a database precluded patenting of innovative disease gene discoveries" (1995).

45. Moreover, in the case of the Mendelian disorders that had dominated human genetics until about this time, the typical gene hunt involved searching for a specific gene (known to play a role in a specific human disease) whose approximate chromosomal location was known (in that it was "near" the flanking maker). In contrast, an EST was an "anonymous" DNA fragment from a random gene of unknown function from an unknown location.

46. Legal restrictions arising outside the patent system, of course, may constrain these managerial privileges.

47. Hilgartner 2009.

48. On the General Public License, see Stallman 2010. See also Kelty 2008 on "recursive publics" and Coleman 2009 on the law and open-source software. On open-source biology, see Hope 2008.

49. Another good example of the remaking of agents are the changes in the US research university since passage of the Bayh–Dole Act in 1980. Although patenting by US universities was not new (see, e.g., Metlay 2006 and Shapin 2008), the Bayh–Dole Act encouraged a large increase in university patent applications in the 1980s and 1990s, generating ongoing debate about what kind of owners universities should be and how they should deploy their intellectual property rights. See, for example, Kenney 1988; Thackray 1998; Krimsky 2003; Slaughter and Rhoads 2004; and Mirowski 2011.

50. Wheeler 1991, A28; Healy 1992.

51. As Adler put it in an article in *Science*, "Like scientists, patent practitioners … formulate and test hypotheses about how patentability changes over time. … Each transaction with the PTO and each judicial patent decision represents an experiment that generates data about the patent system" (1992, 909).

52. To obtain maximum rights, intellectual-property lawyers often draft patent applications with multiple claims organized in a hierarchy of increasing scope, expecting that the PTO might deny some of the far-reaching claims while allowing the narrower claims. The idea is that it is better to let the PTO deny a claim than to prejudge the issue yourself; if necessary, claims can be scaled back during the patent examination process. For an analysis of the rhetoric of a patent application, see Myers 1990.

53. When the NIH revised its first application claiming an additional 2,375 ESTs, it dropped the claims on the proteins (which even some scientists who supported the patent claims believed to be dubious) (Gillis 1992).

54. For comparisons of US and European approaches to biotechnology patents, see Jasanoff 2005, chapter 8, and Parthasarathy 2011.

55. TREVOR, a SEQUENCER, interview, 1992.

56. MARY, a UK FUNDING-AGENCY OFFICIAL, interview, 1992.

57. Trevor, a sequencer, interview, 1992.

58. See, e.g., Stout 1992; Sulston and Ferry 2002, 90.

59. DAVID, an MRC OFFICIAL, interview, November 1992. At the time, the US PTO had rejected the NIH application, but the prospect of an appeal loomed.

60. C. Anderson 1994. Venter describes how he told Varmus that he and coinventor Mark Adams would leave any EST patent royalties to a charity, the NIH Children's Inn (2007, 175–176). Varmus went on to promote access to other "research tools" (see NRC 1997; Winickoff 2015).

61. Kowalski 2000.

62. Fortun 2008. See also Sunder Rajan 2006.

63. C. Anderson 1993.

64. Cook-Deegan 1994b, 55. One of the eight companies discussed in the report, Collaborative Research, Inc., was founded in 1961, so I have not included it in table 5.1. In addition to these eight companies, startups focused on instrumentation for genome research were also founded. I do not discuss them here.

65. HGS and Incyte adjusted their business models when ESTs grew less important.

66. See Sunder Rajan 2006 for a discussion of Incyte and CEO Randy Scott's public performances.

67. Bishop 1994c.

68. Venter's autobiography (Venter 2007) describes how several venture capitalists approached him and offers an account of negotiations with Wallace Steinberg, the head of HealthCare Ventures.

69. NIH 1992.

70. TIGR 1992.

71. HGS 1992.

72. TIGR 1992.

73. Moving from one location to another typically disrupts a laboratory leader's work, but Venter chose to set up TIGR in Maryland near the NIH to avoid losing personnel (2007, 160–161).

74. HGS president William Haseltine told *Science* in October 1994 that HGS and TIGR had 80 sequencing machines between them (Marshall 1994a).

75. *Wall Street Journal* 1993.

76. Moore 1993.

77. HGS 1993 Form 10-K. The US Security and Exchange Commission (SEC) requires publicly traded companies to annually submit this form, which provides a financial report on the company as well as other information on its operations and prospects. These forms are public documents, and are available online from the SEC at https://www.sec.gov.

78. See, for example, Franklin 2003; Sunder Rajan 2006; Helmreich 2008.

79. TIGR's central role in the HGS business model is underlined by the fact that Venter's nonprofit is mentioned in the very first substantive paragraph of the company's 1993 Form 10-K.

80. A specific exception was made for "inventions and patent rights arising out of research funded by certain governmental and not-for-profit organizations" (HGS 1993 Form 10-K).

81. In its 1993 Form 10K, HGS (a.k.a. "the Company") pointed out that "TIGR is independent of the Company and the Company does not have the right to control or direct TIGR's activities."

82. HGS 1993 Form 10-K. In his autobiography, Venter writes that to appease SmithKline, Wallace Steinberg, who had been conducting secret negotiations with SmithKline without Venter's knowledge, pressed for a delay of eighteen months rather than the six months to which he and Venter had originally agreed (Venter 2007, 169, 172).

83. Cook-Deegan 1994b.

84. See HGS 1993 Form 10-K. The agreement was signed in May 1993 and amended in August 1993 and January 1994; HGS later established a collaboration with the company Genetic Therapy.

85. Tied votes would be broken by senior management of both companies or by binding arbitration.

86. HGS 1993 Form 10-K.

87. Boguski 1995.

88. See, for example, Bishop 1994a, 1994b; HGS 1994a; HGS 1994c; Waldholtz 1994; Service 1994.

89. HGS 1994c; see also Venter 2007, 177–178.

90. Papadopoulos, Nicolaides, Wei, et al. 1994. The paper was submitted on February 22, a few days before HGS announced the agreement with Hopkins.

91. Bishop 1994b.

92. HGP scientists regarded the identification of the gene using ESTs as a convincing demonstration of the value of EST databases in gene hunting (field notes, Cold Spring Harbor meeting, May 1994).

93. The sense that many valuable genes might soon be isolated and patented was enhanced by the identification of a long-sought gene involved in breast cancer, BRCA1, about a month after the Vogelstein paper appeared. See Nowak 1994 and Parthasarathy 2007. HGS and TIGR were not involved in the BRCA1 discovery.

94. HGS 1994b.

95. On obligatory passage points, see Latour 1987.

96. Venter defended the licensing arrangement (2007, 181–187).

97. Interview, 2000.

98. On Myriad's patents and business strategies, see Parthasarathy 2007 and Löwy and Gaudillière 2008.

99. Myriad Genetics obtained a second patent on BRCA2, another breast cancer gene. Litigation about the patentability of human genes ultimately went all the way to the US Supreme Court, which ruled against Myriad in 2013 (*Association for Molecular Pathology v. Myriad Genetics*, 569 US [2013]).

100. KEVIN, interview, 1994.

101. Such effects are familiar in economic sociology. Mark Harvey and Andrew McMeekin (2007) make a similar point.

102. NCHGR director Francis Collins was among those who spoke out against using the HGS ESTs in HGP work (Marshall 1994b).

103. Private nonprofits, such as the Wellcome Charitable Trust, were already funding the production of data for public release. Another important case during the HGP was the SNP Consortium, in which a group of pharmaceutical companies supported the production and publication of single-nucleotide polymorphisms, SNPs or "snips."

104. The Columbia libraries were "normalized" to reduce redundancy. Even so, normalized libraries had much duplication (Boguski and Schuler 1995).

105. An earlier incarnation of dbEST had been set up while Venter was still at the NIH. It had been described at the Cold Spring Harbor meeting in 1992. The National Center for Biotechnology Information also sent the data to GenBank's international counterparts, the EMBL Data Library, and the DNA Data Bank of Japan.

106. Boguski 1995; Merck & Company 1995.

107. Merck & Company 1995.

108. As HGS restrictions on TIGR's ESTs began to expire, Venter began transferring them to GenBank; see Adams, Kerlavage, Fleischmann, et al. 1995.

109. Fleischmann, Adams, White, et al. 1995. Only viruses had been completely sequenced previously.

110. TIGR 1997.

111. LEW, interview, 1996.

Chapter 6

1. See Strasser 2008, 2010, 2011; Stevens 2013; and Lenoir 1999. See also Goad 1983.

2. A few laboratories focused on improving sequencing methods; a good example is Leroy Hood's laboratory at Caltech, which developed the automated four-color florescent sequencing machines. For a history of sequencing, see García-Sancho 2012.

3. Bruno Strasser notes that in 1976 ten papers reported nucleic acid sequences; by 1979, the number had grown to 100 (2011, 65).

4. Margaret Oakley Dayhoff, sometimes referred to as the founder of bioinformatics, established the first biomolecular database—namely, a database of protein sequences (see Strasser 2010, 2011; Stevens 2013).

5. On June 30, 1982, the NIH signed a contract with the Los Alamos National Laboratory and Bolt, Beranek and Newman to establish a "public and free nucleic acid sequence databank, which would soon be called GenBank" (Strasser 2008, 538). Strasser (2011) examines the policy-making and grant-review process that led to funding from the NIH, the National Science Foundation, the Department of Agriculture, the DOE, and the Department of Defense. Goad's group was chosen over that of Margaret Dayhoff, who had launched the protein sequence database even before Goad started LASL. Strasser shows that one reason for this decision is that Goad, unlike Dayhoff, did not seek a proprietary interest in the data.

6. Cinkosky, Nelson, and Marr 1988, 1.

7. Fickett, Goad, and Kanehisa 1982, 3–4.

8. Anyone could request online access by mail (Fickett, Goad, and Kanehisa 1982). A variety of academic and commercial entities began providing tools for analyzing the data. Some commercial packages worked with magnetic media that GenBank distributed; others reformatted GenBank data onto their own diskettes (Bilofsky, Burks, Fickett, et al. 1986, 4).

9. Michael Cinkosky, Debra Nelson, and Thomas Marr describe the steps of data in the era of staff-driven collecting: "(1) the data was published; (2) the database staff ... would regularly scan through most of the journals known to publish sequence data ... ; (3) the sequences would then be typed in manually; (4) members of the database staff would then read the original paper to extract the annotation ... ; (5) the completed entry would then go through an extensive review process ... by staff reviewers; (6) when the entry was considered to be satisfactory it would be submitted ... ; (7) sometime later ... the entry would be added to the database; (8) finally, the data would be distributed to end users" (1988, 1).

10. Burks, Cinkosky, Gilna, et al. 1990, esp. 8–14.

11. BRIAN, an architect of GenBank, interview, 1991.

12. See, for example, Kabat 1989.

13. DAVE, interview, 1992.

14. For a sense of this crisis mood, see, for example, Colwell, Swartz, MacDonnell, et al. 1989, a volume titled *Biomolecular Data: A Resource in Transition*, which came out of a workshop that was planned throughout 1986 and held in May 1987.

15. Methods are described succinctly; materials that cannot be rendered on the printed page may not be presented at all; and data are cleaned up, analyzed, summarized, and concisely packaged—a process that continues in postpublication texts, such as literature reviews and news accounts (Latour 1987).

16. See Cinkosky, Fickett, Gilna, et al. 1992.

17. Walker 1989, 46–47.

18. Ibid., 46–47.

19. Ibid.

20. Wells 1989, 337.

21. By the mid-1980s, GenBank had concluded that in-house abstracting was incompatible with the exponentially increasing rate of sequencing (see Burks 1987; Wells 1989; Burks, Cinkosky, Gilna, et al. 1990, 11; and Strasser 2011).

22. A report on LASL envisioned "encouraging direct contributions by authors," and as early as 1983 GenBank began "trying to get support from the journals to encourage or enforce" direct submission (Fickett, Goad, and Kanehisa 1982, 5; see also Kanehisa, Fickett, and Goad 1982, 157; Goad 1983; Lenoir 1999; Strasser 2011).

23. Bilofsky, Burks, Fickett, et al. 1986, 3; Burks 1987, 5.

24. EMBL Data Library and GenBank Staff 1987. "A new system for direct submission of data to the nucleotide sequence banks" (EMBL Data Library and GenBank Staff 1987, n.p.).

25. Walker 1989, 48.

26. Changes in software included shifting from flat files to a relational database system.

27. G. Bell 1988, 8–9.

28. IntelliGenetics, which took over the GenBank contract from Bolt, Beranek and Newman in 1987, developed the AUTHORIN software (see Lenoir 1999). See also Burks, Cinkosky, Gilna, et al. 1990. According to one of the GenBank principals, by

the end of 1991 more than 60 percent of submissions were entered using AUTHORIN.

29. GenBank 1993.

30. *NCBI News* 1995. On the rise of the Internet and the World Wide Web, see Abbate 1999.

31. CHRIS, a US FUNDING-AGENCY OFFICIAL, interview, 1994.

32. Ibid.

33. EMBL Data Library and GenBank Staff 1987, n.p. The *Proceedings of the National Academy of Sciences* put into effect a similar direct-submission policy on January 1, 1989. See Dawid 1989, 407, on its policy.

34. EMBL Data Library and GenBank Staff 1987, n.p., emphasis in original.

35. McCain 1995 surveys journal policies.

36. Staff-driven collecting continued alongside direct submission for several years, first at Los Alamos and then at the National Center for Biotechnology Information (NCBI) after GenBank moved to NCBI in 1992. NCBI was created in 1988 and made part of the National Library of Medicine. After GenBank moved to NCBI, it piggy-backed on the National Library of Medicine's indexing service Medline to identify scientific papers containing sequences that otherwise might have been missed (MARK, NCBI, interview, 1993).

37. Maddox 1989b, 277.

38. Cameron, Kahn, and Philipson 1989, 848.

39. R. Roberts 1989, 114.

40. Maddox 1989a, 855. Journals today do enforce compliance with regulations pertaining to human subjects, and controversy surrounds journals' responsibilities regarding secrecy versus publication of information relevant to pathogens and bioweapons (on pathogens, see Vogel 2013).

41. The new policy took effect in 1996.

42. R. Roberts 1989, 114. Commenting on the dispute, Robert Cook-Deegan compared "journal editors kneeling before the alter of the printed page, and the printed page alone" to "dinosaurs" doomed for extinction (1994a, 289).

43. Cinkosky, Fickett, Gilna, et al. 1991, 1273.

44. Cinkosky, Fickett, Gilna, et al. 1992.

45. Cinkosky, Fickett, Gilna, et al. 1991, 1276. This policy resulted in some compli-cations for GenBank—for example, in verifying that a paper had been published, which made it appropriate to publicly release a previously confidential sequence.

46. Cinkosky, Fickett, Gilna, et al. 1992, 271.

47. Cinkosky, Fickett, Gilna, et al. 1991, 1276.

48. Brian, an architect of GenBank, interview, 1991.

49. George Orwell, *1984*.

50. During this period, the nucleic acid sequence databases and the protein sequence databases followed different policies. Protein sequence databases were committed to the "dynamic maintenance of an integrated set of information reflecting the current biological understanding of the data," whereas the nucleic acid databases maintained a repository of submitted and published information (Mewes, Doelz, and George 1994, 249).

51. Cinkosky, Fickett, Gilna, et al. 1991, 1274.

52. Ibid., 1276.

53. See Kohler 1994 on the *Drosophila* geneticists and de Chadarevian 2004 on the worm geneticists; see also Kelty 2012. My interlocutors frequently commented on worm biologists' commitment to openness in science.

54. Field notes, 1990, 1992, 1994. See also Sulston and Ferry 2002.

55. Technician A told me in 1992, "they tell me to sequence something, I sequence it"; "it's a job" (interview, 1992). Another technician stated, "Technicians, as far as I'm concerned, this is myself included, of course, are kind of, they're almost equipment. You don't feel like you're really a part of the project" (Technician B, interview, 1992). Some technicians I met, however, derived considerable satisfaction out of being part of the genome project, and some of them were pleased when their work was recognized by being included on the author list of papers.

56. Field notes, 1992.

57. NCHGR 1992a, tab L.

58. Some regions of DNA could not be sequenced (e.g., owing to repeated sequences or intractable cloning problems).

59. Because inserting and deleting a base shifts the frame of the three-base triplets that code for specific amino acids, genome researchers consider INsertions and DELetions (referred to as "indels") to be worse than miscalled bases.

60. Jim, NHGRI official, interview, 2000.

61. Hilgartner 1998; Joly and Mangematin 1998.

62. What counted as a major milestone rapidly evolved as sequencing accelerated.

63. Dujon, Alexandraki, André, et al. 1994.

64. Goffeau, Barrell, Bussey, et al. 1996.

65. Sulston and Ferry 2002, 143. I was unable to attend the Bermuda meetings, so this account is based on the published literature, interviews, and conversations with genome scientists.

66. Ibid.

67. Ibid., 143–145.

68. Ibid., 145.

69. Sulston and Waterston were initially alone in doing daily releases. Meeting participants included Craig Venter, who later argued in *Science* against the policy (see Adams and Venter 1996). Others agreed with the goal but disagreed about implementation (see Marshall 1996). One center wanted to hold data for more than a day but less than three months, and some objected that numerous small submissions would burden the databases (Marshall 1996). Some scientists from the smaller US sequencing laboratories told me that the larger players had imposed policy on them. On the construction of consensus in science advisory groups, see Gilbert and Mulkay 1984 and Hilgartner 2000.

70. When the principles were reaffirmed in 1997, the size constraint was adjusted to 2 kilobases (2,000 bases); sequences that large or larger were deemed "usable" for searching the databases for similar sequences and other types of sequence analysis. See Ouellette and Boguski 1997, 953.

71. HUGO 1996.

72. See Contreras 2011 for a legal scholar's analysis of the Bermuda Principles using Elinor Ostrom's concept of common-property regimes.

73. Jim, NHGRI official, interview, 2000.

74. Adams and Venter 1996, 535.

75. Bentley 1996, 533.

76. Quoted in Ouellette and Boguski 1997, 954.

77. Adams and Venter 1996, 535, 536.

78. Germany, which was not making the large investment being made by the United States and the United Kingdom, succumbed to "international" pressure and joined in rapid release (Alison Abbott 1997).

79. HUGO 1997; Guyer 1998. The Bermuda Principles also had an ongoing legacy in that rapid data release was required in a number of genomics and proteomics projects undertaken after the HGP was completed. For a review of these projects, see Contreras 2011.

80. Two other groups, Glen Evans's and Claire Frazer's, also participated.

81. Such research was of interest for several reasons, including that the mouse is used as a model for the human in biomedical research.

82. The MHC episode was not unique; disputes involving projects to sequence the genome of plasmodium, the parasite that causes malaria, occurred at about the same time (see, e.g., Gottlieb, McGovern, Goodwin, et al. 2000; Macilwain 2000; Pace 2001; Roos 2001).

83. Hyman 2001, 827.

84. David Roos argued that "the identification of individual genes of interest for further experimental analysis must be acceptable—perhaps even without the need for formal permission—otherwise, early data release serves no purpose at all. Conversely, second-party publication of raw, unpublished, sequence data posted on the Web must be viewed as violating ethical standards—analogous to the verbatim plagiarism of unpublished results from a meeting presentation" (2001, 1261).

85. NHGRI 2000.

86. SHEILA, a SEQUENCER, interview, May 2001. Also, RANDY, a journal editor, commented, "We're in centers to develop careers as biologists. ... If you make it sound like we're flipping burgers at McDonald's, we're going to lose good people, and ... the field will stagnate" (interview, May 2001).

87. The language in quotation marks is from three interviews (Randy, Jim, and Sheila) conducted in May 2001.

88. Jim, NHGRI official, interview, 2001.

89. Randy, a journal editor, interview, May 2001.

90. Timing was a point of contention in the exchange between Rowen (and her colleagues) and Elaine Bell, editor of *Immunology Today* (see E. Bell 2000; Rowen and Hood 2000; Rowen, Wong, Lane, et al. 2000).

91. *Sharing Data* (Fort Lauderdale Agreement) 2003, 2.

92. On proteomics, see, for example, McNally and Glasner 2007; see also Davies, Frow, and Leonelli 2013 on other "big biology" projects.

93. *Sharing Data* (Fort Lauderdale Agreement) 2003, 2–4, emphasis added.

94. *Sharing Data* (Fort Lauderdale Agreement) 2003, 3–4.

95. Ibid., 4.

96. Ibid., 4.

97. NHGRI 2003.

98. This particular settlement provided a framework for launching new projects, but as one might expect from the account given in this chapter, it by no means brought permanent stability, and new CRPs generated debates about the allocation of control. See Contreras 2011 for a review of several post-HGP examples.

Chapter 7

1. See Rödder, Franzen, and Weingart 2012 on the "medialization" of science in the contemporary world.

2. Bruce, the chair of a key US advisory committee, for example, confidently asserted that one of these approaches, probably hybridization arrays, was bound to work. One notable exception was a scientist who had helped develop the four-color automated machines. In 1989, he assured me that the genome would be sequenced using the four-color technology. Incremental improvements would be enough, he said, whereas completely new sequencing technologies would take too long to develop.

3. In 1993 and 1994, a number of knowledgeable sequencers told me that people were abandoning hope that any of the novel methods would be ready in time. Some described a meeting at NIH as a key moment because no one working on any of the novel technologies had been willing to assert that the technology would be ready in time.

4. Marshall 1995; see also Sulston and Ferry 2002, 114–139.

5. In October 1995, Maynard Olson published a position paper in *Science* supporting the shift to sequencing.

6. Lew, Coastal Center director, interview, 1996.

7. NCHGR 1995.

8. Hughes 1983.

9. Ollie, interview, 1996. Artificial-intelligence methods would not be required. "Old-fashioned, brute-force statistical techniques" could do it by taking into account more variables and measuring more things. One problem without a technical solution, Ollie argued, was the cost of the patented reagents needed to use the automated sequencers because if one reduces the quantity of reagents needed, the company with the intellectual-property rights can simply increase the unit price.

10. One prominent example was Sulston (see Sulston and Ferry 2002, 119).

11. Confidential memorandum, May 1994.

12. Ollie, interview, January 1996.

13. Ibid.

14. Any region of the genome for which a contig map has been produced can be sequenced using the clone-by-clone approach. It is therefore not necessary to delay sequencing until after a map of an entire genome has been constructed.

15. Fleischmann, Adams, White, et al. 1995. Venter's group was the first to successfully implement the WGS method for a bacterial genome. Several people had previously proposed the concept (e.g., Fred Blattner early in the HGP).

16. See Bostanci 2004 for further analysis of the clone-by-clone strategy versus the WGS strategy as experimental traditions.

17. Weber and Myers 1997.

18. Green 1997.

19. See Hilgartner 2012 for a more detailed account of the period following the announcement.

20. Venter 2007. Setting up a new company, as opposed to having Perkin-Elmer undertake the project itself, had numerous advantages, not least of which were limiting Perkin-Elmer's exposure, establishing a new entity that could raise capital by making an IPO, and creating attractive positions for top managers.

21. A scientist with knowledge of company decision making told me that the New Company's decision about when to unveil its plan to the media was made for "lots of reasons": several investment firm meetings were coming up; the Cold Spring Harbor meeting offered a chance to explain the plan to the genome community; rumors were starting to circulate; and the company needed to announce its new sequencing machine because word of the new machine was getting out, which threatened to hurt sales of its existing sequencing machines (interview, May 1998).

22. Sulston and Ferry 2002, 149–152; Venter 2007, 238–243.

23. Venter 2007, 241. The press release, dated May 9, was loaded into the Business Wire press release distribution service on Sunday, May 10.

24. Wade 1998c.

25. Ibid.

26. Ibid.

27. Perkin-Elmer and TIGR 1998.

28. On "story stocks" in genomics, see Fortun 2001. To conform with securities law obliging Perkin-Elmer (a company listed on the New York Stock Exchange) to disclose major developments and risks, the press release included a standard disclosure paragraph, noting that "certain statements in this press release and its attachments are forward-looking" and that "a variety of factors" could cause "actual results" to

differ from "anticipated results" (Perkin-Elmer and TIGR 1998). Such disclosures as well as the promissory and speculative nature of genomics have attracted much comment (e.g., Fortun 2001, 2008; Rabinow and Dan-Cohen 2005; Sunder Rajan 2006; see also Hedgecoe 2004 and Tutton 2011).

29. Perkin-Elmer 1998b. The press release was also dated May 9 and loaded into the Business Wire press release distribution service on Sunday, May 10.

30. Perkin-Elmer and TIGR 1998.

31. Sulston and Ferry 2002, 152.

32. Interview, May 1998.

33. Gillis and Weiss 1998. In addition, an Associated Press story, "Gene Technology Must Be Proven Before Government Will Use It," led with the idea that the new technological strategy "will have to prove itself" before the HGP will adopt it (Recer 1998).

34. Wade 1998a.

35. JOHN, interview, May 1998.

36. Sulston and Ferry 2002, 153–158; Venter 2007, 243–245.

37. Sulston and Ferry 2002, 162.

38. Ibid.

39. Wellcome Trust 1998.

40. Bruce Lewenstein (1995) describes similar information gathering in his analysis of the cold-fusion episode.

41. On Myriad and its breast cancer gene patents, see Parthasarathy 2007 and Löwy and Gaudillière 2008.

42. Recer 1998.

43. On the politics of "research tools," see Winickoff 2015; see also NRC 1997.

44. Some of these scientists mentioned the published critique by Philip Green (1997).

45. KEVIN, interview, May 1998.

46. Ibid.

47. Irwin 1998.

48. Wade 1998b.

49. Patrinos discussed this issue in his answers to questions submitted by members of Congress after the hearing (US House of Representatives 1998, 88–89).

50. Sulston and Ferry 2002, 164.

51. Collins, Patrinos, Jordan, et al. 1998, 683. See also Waterston and Sulston 1998, 53–54.

52. Collins, Patrinos, Jordan, et al. 1998, 683.

53. NHGRI 1998. Bostanci 2010 shows that the metaphors of a "rough draft" of the sequence and a "draft" version of "the Book of Life" were new to the language of the HGP, suggesting that this concept developed only as the HGP responded to Celera.

54. Shreeve 2004; Sulston and Ferry 2004; Venter 2007. Shreeve was given access to Celera, and he writes that his request to the NHGRI for similar access was denied (2004, 375–376).

55. William Blair & Company 1999, 22, 2, emphasis in original.

56. On the concept of "public identity," see Hilgartner 2000, 43–44.

57. Perkin-Elmer 1998a.

58. Peter, interview, May 2000.

59. William Blair & Company 1999.

60. Celera had signed a contract to exclusively license therapeutic proteins covered by the agreement to Pfizer, with Celera receiving milestone payments and royalties (ibid.).

61. Peter, interview, May 2000.

62. The common-heritage view, expressed in one way or another by many HGP scientists during the project, is a theme in John Sulston's account (Sulston and Ferry 2002). NCHGR director Francis Collins, a devout Christian, referred to the sequence as "the language of God," "God's instruction book," and "our shared inheritance" (2006, title, 109, 121).

63. Hilgartner 2007. For another treatment of Venter's persona, see Shapin 2008.

64. Belkin 1998, 28.

65. In a retrospective account, Maynard Olson (2002) describes how in testimony before Congress Venter had denied that the WGS strategy would produce a "quick-and-dirty" sequence, saying that he expected the company's sequence to be better than what the HGP planned to produce. See also US House of Representatives 1998.

66. Olson argued that as of December 2001 a coalition of publicly funded laboratories had been working for a year and a half "to straighten out Celera's fruit-fly sequence and this effort remains well short of the goal. Until the clean-up job is

complete, there will be no meaningful basis on which to assess how well Celera's methods worked even on the fruit fly" (2002, 939).

67. See Rödder 2009 and Schäfer 2009 for analyses of the coverage of these milestones.

68. Collins, Patrinos, Jordan, et al. 1998. Regarding long-range contiguity, the HGP stated in 1998: "Although production of contiguous sequence without gaps is the goal, any irreducible gaps must be annotated as to size and position" (ibid., 684). As discussed in chapter 6, sequencing errors can consist of miscalled bases or indels (insertions or deletions of bases), which are more troublesome.

69. Ezrahi 1990.

70. See Jasanoff 1996.

71. Clinton 2000b.

72. On this imaginary, see Hilgartner 2015, 36–37.

73. Clinton 2000a.

74. Ibid.

75. Ibid.

76. See also Hurlbut 2015 and Hilgartner, Prainsack, and Hurlbut in press.

77. International Human Genome Sequencing Consortium 2001; Venter, Adams, Myers, et al. 2001.

78. NRC 2003b.

79. For example, F. Collins 2001, 641.

80. Kevin, interview, May 2001.

81. William Blair & Company 1999.

82. The matter was debated in a series of papers: Myers, Sutton, Smith, et al. 2002; Waterston, Lander, and Sulston 2002. The issue was not just that Celera used the HGP data, which the company readily acknowledged. The question was whether in addition to the raw, unordered information in the HGP reads, Celera's assembly method retained information about genomic order that was embedded in the HGP data. Waterston, Lander, and Sulston concluded: "The data (with a majority of the underlying sequence information and all mapping information coming from the HGP), the methodology (with approaches that preserve [HGP] assembly information), and the properties of the assembly (extensive sequence continuity) … all indicate that the assembly reported in the Celera article is instead appropriately viewed as a refinement built on the HGP assemblies" (2003, 3024). The debate is too technical to unpack here.

83. Venter 2007. See also Rabinow and Dan-Cohen 2005 on Celera Diagnostics.

84. Jim, interview, July 2000.

85. Collins, Morgan, and Patrinos 2003; NRC 2003a.

Chapter 8

1. Calvert 2008; Carlson 2010; Richardson and Stevens 2015.

2. F. Collins 2010, 675.

3. Indeed, one might argue that the "superorganisms" of high-energy physics aim to achieve something like this (see Knorr Cetina 1999).

4. Outside of the US regime's jurisdiction, the successes of the Sanger Centre and Généthon were also especially important.

Appendix

1. Jason, interview, May 1990.

2. On a few occasions, I conducted simultaneous interviews with more than one interlocutor. Also, some interviews took place in restaurants or other spaces that may have allowed bystanders some access to what was said.

3. For an ethnographic examination of the multivalent nature of informed consent in social research, see the special issue of *Political and Legal Anthropology Review* titled "New Bureaucracies of Virtue" (Jacob and Riles 2007).

4. Dylan, interview, 1996.

5. Jacob, interview, 1993.

6. Clark, interview, 1993.

7. Nicole, interview, 1992.

8. Interview, Tim, 1995.

9. Kasia, interview, 1992.

10. Ibid.

11. Field notes, 1990.

12. Goffman 1959.

13. The meetings were later retitled "Genome Sequencing and Biology" to reflect the completion of the HGP's mapping phase.

14. The rule was known as the "Inglefinger rule," after the editor who developed it. For a discussion, see Nelkin 1987.

15. In 2003, my house burned down, and virtually all of my family's belongings were destroyed. Some of my field notes, especially those in handwritten rather than electronic form, were among the casualties. Other handwritten notes and paper documents experienced damage from fire and water but were salvageable. I cannot locate the handwritten notes taken at this meeting, which I believe were destroyed in the fire, so this account is based on my recollection of the event, not on edited field notes. This description of the conversation between Montague and Harold is the only instance in the book in which an account was wholly constructed from memory.

References

Abbate, Janet. 1999. *Inventing the Internet.* Cambridge, MA: MIT Press.

Abbott, Alison. 1997. Germany Rejects Genome Data "Isolation." *Nature* 387:536.

Abbott, Andrew. 1988. *The System of Professions: An Essay on the Division of Expert Labor.* Chicago: University of Chicago Press.

Adams, Mark D., Jenny M. Kelley, Jeannine D. Gocayne, Mark Dubnik, Michael H. Polymeropoulous, Hong Xiao, Carl R. Merril, et al. 1991. Complementary DNA Sequencing: Expressed Sequence Tags and the Human Genome Project. *Science* 252 (5013): 1651–1656.

Adams, Mark D., Mark Dubnick, Anthony R. Kerlavage, Ruben Moreno, Jenny M. Kelley, Teresa R. Utterback, James W. Nagle, Chris Fields, and J. Craig Venter. 1992. Sequence Identification of 2,375 Human Brain Genes. *Nature* 355: 632–634.

Adams, Mark D., Anthony R. Kerlavage, Robert D. Fleischmann, Rebecca A. Fuldner, Carol J. Bult, Norman H. Lee, Ewen F. Kirkness, et al. 1995. Initial Assessment of Human Gene Diversity and Expression Patterns Based upon 83 Million Nucleotides of cDNA Sequence. *Nature* 377 (6547 Suppl.): 3–174.

Adams, Mark D., and J. Craig Venter. 1996. Should Non-Peer-Reviewed Raw DNA Sequence Data Release Be Forced on the Scientific Community? *Science* 274 (5287): 534–536.

Adler, Reid G. 1992. Genome Research: Fulfilling the Public's Expectations for Knowledge and Commercialization. *Science* 257 (5072): 908–914.

Anderson, Benedict. 1991. *Imagined Communities: Reflections on the Origin and Spread of Nationalism.* New York: Verso.

Anderson, Christopher. 1993. Genome Project Goes Commercial. *Science* 259 (5093): 300–302.

Anderson, Christopher. 1994. NIH Drops Bid for Gene Patents. *Science* 263 (5149): 909.

Ankeny, R. A., and S. Leonelli. 2011. What Is so Special about Model Organisms? *Studies in the History and the Philosophy of Science: Part A* 42 (2): 313–323.

Atkinson, Paul, Claire Batchelor, and Evelyn Parsons. 1998. Trajectories of Collaboration and Competition in a Medical Discovery. *Science, Technology & Human Values* 23 (3): 259–284.

Austin, J. L. [1962] 1975. *How to Do Things with Words*. 2nd ed. Cambridge, MA: Harvard University Press.

Balmer, Brian. 1996a. Managing Mapping in the Human Genome Project. *Social Studies of Science* 26 (3): 531–573.

Balmer, Brian. 1996b. The Political Cartography of the Human Genome Project. *Perspectives on Science* 4 (3): 249–282.

Balmer, Brian. 1998. Transitional Science and the Human Genome Mapping Project Resource Center. In *Genetic Imaginations: Ethical, Legal, and Social Issues in Human Genome Research*, ed. Peter E. Glasner and Harry Rothman, 7–19. Aldershot, UK: Ashgate.

Balmer, Brian. 2006. A Secret Formula, a Rogue Patent, and Public Knowledge about Nerve Gas: Secrecy as a Spatial-Epistemic Tool. *Social Studies of Science* 36 (5): 691–722.

Balmer, Brian, Rosemary Davidson, and Norma Morris. 1998. Funding Research through Directed Programmes: AIDS and the Human Genome Project in the UK. *Science & Public Policy* 25 (3): 185–194.

Baltimore, David. 1987. Genome Sequencing: A Small-Science Approach. *Issues in Science and Technology* 3:48–50.

Barnes, Barry, and John Dupré. 2008. *Genomes and What to Make of Them*. Chicago: University of Chicago Press.

Barry, Andrew. 2006. Technological Zones. *European Journal of Social Theory* 9 (2): 239–253.

Beckwith, Jon. 2002. *Making Genes, Making Waves: A Social Activist in Science*. Cambridge, MA: Harvard University Press.

Belkin, Lisa. 1998. DNA Is His Pay Dirt. *New York Times Magazine*, August 23.

Bell, Elaine. 2000. Publication Rights for Sequence Data Producers (letter to the editor). *Science* 290 (5497): 1696–1698.

Bell, George I. 1988. The Human Genome—Computational Challenges. In *Proceedings for the Third International Conference on Supercomputing*, 1–18. Los Alamos, NM: Los Alamos National Laboratory.

Bentley, David R. 1996. Genomic Sequence Information Should Be Released Immediately and Freely in the Public Domain. *Science* 274 (5287): 533–534.

Biagioli, Mario, and Peter Galison, eds. 2003. *Scientific Authorship: Credit and Intellectual Property in Science*. London: Routledge.

Bilofsky, Howard, Christian Burks, James W. Fickett, Walter B. Goad, Frances I. Lewitter, Wayne P. Rindone, C. David Swindell, and Chang-Shung Tung. 1986. The GenBank Genetic Sequence Databank. *Nucleic Acids Research* 14 (1): 1–4.

Bishop, Jerry E. 1994a. Another Gene Found to Cause Colon Cancer. *Wall Street Journal*, March 17.

Bishop, Jerry E. 1994b. Human Genome Sciences Sees Gold in Trove of Genes—Colon-Cancer Report Spotlights Strategy in Exploiting Biggest Collection. *Wall Street Journal*, March 18.

Bishop, Jerry E. 1994c. Secret Science: The Ability to Patent Genetic Information May Be Hampering Researchers' Efforts. *Wall Street Journal*, May 20.

Boguski, Mark S. 1995. The Turning Point in Genome Research. *Trends in Biochemical Sciences* 20 (8): 295–296.

Boguski, Mark S., and Gregory D. Schuler. 1995. ESTablishing a Human Transcript Map. *Nature Genetics* 10 (4): 369–371.

Bostanci, Adam. 2004. Sequencing Human Genomes. In *From Molecular Genetics to Genomics: The Mapping Cultures of Twentieth-Century Genetics*, ed. Jean-Paul Gaudillière and Hans-Jörg Rheinberger, 158–179. New York: Routledge.

Bostanci, Adam. 2010. A Metaphor Made in Public. *Science Communication* 32 (4): 467–488.

Boyle, James. 1996. *Shamans, Software, and Spleens: Law and the Construction of the Information Society*. Cambridge, MA: Harvard University Press.

Boyle, James. 2003. Enclosing the Genome: What the Squabbles over Genetic Patents Can Teach Us. *Advances in Genetics* 50:97–122.

Brannigan, Augustine. 1981. *The Social Basis of Scientific Discoveries*. New York: Cambridge University Press.

Brenner, Sydney. 1990. The Human Genome: The Nature of the Enterprise. In *Human Genetic Information: Science, Law and Ethics*, Ciba Foundation Symposium no. 149, 6–17. Chichester, UK: John Wiley & Sons.

Brown, Nik, Brian Rappert, and Andrew Webster, eds. 2000. *Contested Futures: A Sociology of Prospective Techno-Science*. Aldershot, UK: Ashgate.

Burks, Christian. 1987. The GenBank Database and the Flow of Sequence Data for the Human Genome. Paper presented at Brookhaven National Laboratory Symposium on the Human Genome, September 14-16.

Burks, Christian, Michael J. Cinkosky, Paul Gilna, Jamie E.-D. Hayden, Yuki Abe, Edwin J. Atencio, Steve Barnhouse, et al. 1990. GenBank: Current Status and Future Directions. *Methods in Enzymology* 183:3–22.

Callon, Michel. 1994. Is Science a Public Good? Fifth Mullins Lecture, Virginia Polytechnic Institute, 23 March 1993. *Science, Technology & Human Values* 19 (4): 395–424.

Calvert, Jane. 2008. The Commodification of Emergence: Systems Biology, Synthetic Biology, and Intellectual Property. *Biosocieties* 3 (4): 383–398.

Cambrosio, Alberto, and Peter Keating. 1995. *Exquisite Specificity: The Monoclonal Antibody Revolution.* New York: Oxford University Press.

Cameron, Graham, Patricia Kahn, and Lennart Philipson. 1989. Journals and Databanks (letter to the editor). *Nature* 342 (6252): 848.

Campbell, Brian. 1985. Uncertainty as Symbolic Action in Disputes among Experts. *Social Studies of Science* 15 (3): 429–453.

Campbell, Eric G., Brian R. Clarridge, Manjusha Gokhale, Lauren Birenbaum, Stephen Hilgartner, Neil A. Hotlzman, and David Blumenthal. 2002. Data Withholding in Academic Genetics: Evidence from a National Survey. *Journal of the American Medical Association* 287 (4): 473–480.

Carlson, Robert H. 2010. *Biology Is Technology: The Promise, Peril, and New Business of Engineering Life.* Cambridge, MA: Harvard University Press.

Carruthers, Bruce G., and Laura Ariovich. 2004. The Sociology of Property Rights. *Annual Review of Sociology* 30:23–46.

Christensen, Clayton. 1997. *The Innovator's Dilemma: When New Technologies Cause Great Firms to Fail.* Cambridge, MA: Harvard Business School Press.

Church, George M., and Stephen Kieffer-Higgins. 1988. Multiplex DNA Sequencing. *Science* 240 (4849): 185–188.

Cinkosky, Michael J., James W. Fickett, Paul Gilna, and Christian Burks. 1991. Electronic Data Publishing and GenBank. *Science* 252 (5010): 1273–1277.

Cinkosky, Michael J., James W. Fickett, Paul Gilna, and Christian Burks. 1992. Electronic Data Publishing in GenBank. *Los Alamos Science* 20:270–272.

Cinkosky, Michael J., Debra Nelson, and Thomas G. Marr. 1988. *GenBank/HGIR (Human Genome Information Resource) Technical Manual.* Los Alamos, NM: Los Alamos National Laboratory.

Clarke, Adele, and Joan H. Fujimura, eds. 1992. *The Right Tools for the Job: At Work in the Twentieth-Century Life Sciences.* Princeton, NJ: Princeton University Press.

Clinton, William F. 2000a. Remarks Made by the President on the Completion of the First Survey of the Entire Human Genome. Washington, DC, June 26.

Clinton, William F. 2000b. State of the Union Address. Washington, DC, January 27.

Coleman, Gabriella. 2009. Code Is Speech: Legal Tinkering, Expertise, and Protest among Free and Open Source Software Developers. *Cultural Anthropology* 24 (3): 420–454.

Collingridge, David. 1981. *The Social Control of Technology*. New York: St. Martin's Press.

Collins, Francis S. 2001. Contemplating the End of the Beginning. *Genome Research* 11: 641–643.

Collins, Francis S. 2006. *The Language of God: A Scientist Provides Evidence for Belief.* Chicago: University of Chicago Press.

Collins, Francis S. 2010. Has the Revolution Arrived? *Nature* 464 (7289): 674–675.

Collins, Francis S., and David Galas. 1993. A New Five-Year Plan for the U.S. Human Genome Project. *Science* 262 (4346): 43–46.

Collins, Francis S., Michael Morgan, and Aristides Patrinos. 2003. The Human Genome Project: Lessons from Large-Scale Biology. *Science* 300 (5617): 286–290.

Collins, Francis S., Ari Patrinos, Elke Jordan, Aravinda Chakravarti, Raymond Gesteland, LeRoy Walters, and the members of the DOE and NIH planning groups. 1998. New Goals for the U.S. Human Genome Project: 1998–2003. *Science* 282 (5389): 682–689.

Collins, Harry M. 1985. *Changing Order: Replication and Induction in Scientific Practice.* London: Sage.

Colwell, Rita R., David G. Swartz, Michael Terrell MacDonell, and CODATA, eds. 1989. *Biomolecular Data: A Resource in Transition.* Oxford: Oxford University Press.

Contreras, Jorge L. 2011. Bermuda's Legacy: Policy, Patents, and the Design of the Genome Commons. *Minnesota Journal of Law, Science & Technology* 12 (1): 61–125.

Cook-Deegan, Robert M. 1994a. *The Gene Wars: Science, Politics, and the Human Genome.* New York: Norton.

Cook-Deegan, Robert M. 1994b. *Survey of Genome Science Corporations.* Contract Report prepared for the Office of Technology Assessment, U.S. Congress. Washington, DC: US Government Publication Office.

Cox, David, Margin Burmeister, E. Roydon Price, Sywon Kim, and Richard M. Myers. 1990. Radiation Hybrid Mapping: A Somatic Cell Genetic Method for

Constructing High-Resolution Maps of Mammalian Chromosomes. *Science* 250 (4978): 245–250.

Cozzens, Susan E. 1989. *Social Control and Multiple Discovery in Science: The Opiate Receptor Case*. Albany: State University of New York Press.

Daston, Lorraine. 1995. The Moral Economy of Science. *Osiris* 10:2–24.

Davies, Gail, Emma Frow, and Sabina Leonelli. 2013. Bigger, Faster, Better? Rhetorics and Practices of Large-Scale Research in Contemporary Bioscience. *Biosocieties* 8 (4): 386–396.

Davis, Bernard D., and Colleagues. 1990. The Human Genome and Other Initiatives. *Science* 249 (4967): 342–343.

Davis, Kevin. 2001. *Cracking the Genome: Inside the Race to Unlock Human DNA*. New York: Free Press.

Dawid, Igor B. 1989. How to Prepare a Manuscript for PNAS. *Proceedings of the National Academy of Sciences of the United States of America* 86 (1): 407.

Day, George, and Paul Schoemaker, with Robert E. Gunther. 2000. *Wharton on Managing Emerging Technologies*. New York: Wiley.

De Chadarevian, Soraya. 2002. *Designs for Life: Molecular Biology after World War II*. Cambridge: Cambridge University Press.

De Chadarevian, Soraya. 2004. Mapping the Worm's Genome: Tools, Networks, Patronage. In *From Molecular Genetics to Genomics: The Mapping Cultures of Twentieth-Century Genetics*, ed. Jean-Paul Gaudillière and Hans-Jörg Rheinberger, 95–110. New York: Routledge.

Department of Energy (DOE). 1992. *Human Genome: 1991–1992 Program Report*. Report no. DOE ER-0544P, June 1992.

Dennis, Michael Aaron. 1994. "Our First Line of Defense": Two University Laboratories in the Postwar American State. *Isis* 85 (3): 427–455.

Dennis, Michael Aaron. 1999. Secrecy and Science Revisited: From Politics to Historical Practice and Back. In *Secrecy and Knowledge Production*, ed. Judith Reppy, 1–16. Ithaca, NY: Cornell University Peace Studies Program.

Doing, Park. 2009. *Velvet Revolution at the Synchrotron: Biology, Physics, and Change in Science*. Cambridge, MA: MIT Press.

Doggett, Norman A., Raymond L. Stallings, Carl E. Hildebrand, and Robert K. Moyzis. 1992. The Mapping of Chromosome 16. *Los Alamos Science* 20:182–210.

Dujon, B., D. Alexandraki, B. André, W. Ansorge, V. Baladron, J. P. Ballesta, A. Banrevi, et al. 1994. Complete DNA Sequence of Yeast Chromosome XI. *Nature* 369 (6479): 371–378.

Duster, Troy. 1990. *Backdoor to Eugenics*. New York: Routledge.

Edgerton, David. 2007. *The Shock of the Old: Technology and Global History since 1900*. Oxford: Oxford University Press.

Eisenberg, Rebecca S. 1990. Patenting the Human Genome. *Emory Law Journal* 39 (3): 721–745.

Epstein, Steven. 1996. *Impure Science: AIDS, Activism, and the Politics of Knowledge*. Berkeley and Los Angeles: University of California Press.

European Molecular Biology Laboratory (EMBL) Data Library and GenBank Staff. 1987. A New System for Direct Submission of Data to the Nucleotide Sequence Banks. *Nucleic Acids Research* 18 (18): front matter (not paginated).

Evans, James A. 2010. Industry Collaboration, Scientific Sharing, and the Dissemination of Knowledge. *Social Studies of Science* 40 (5): 757–791.

Ezrahi, Yaron. 1990. *The Descent of Icarus: Science and the Transformation of Contemporary Democracy*. Cambridge, MA: Harvard University Press.

Fickett, James. 1989. The Database as a Communication Medium. In *Biomolecular Data: A Resource in Transition*, ed. Rita R. Colwell, David G. Swartz, Michael Terrell MacDonell, and CODATA, 295–302. Oxford: Oxford University Press.

Fickett, James, Walter Goad, and Minoru Kanehisa. 1982. *Los Alamos Sequence Library: A Database and Analysis System for Nucleic Acid Sequences*. Los Alamos, NM: Los Alamos National Laboratory.

Fleck, Ludwik. [1935] 1979. *Genesis and Development of a Scientific Fact*. Trans. F. Bradley and T. J. Trenn. Chicago: University of Chicago Press.

Fleischmann, Robert D., Mark D. Adams, Owen White, Rebecca A. Clayton, Ewen F. Kirkness, Anthony R. Kerlavage, Carol J. Bult, et al. 1995. Whole-Genome Random Sequencing and Assembly of Haemophilus Influenzae Rd. *Science* 269 (5223): 469–512.

Foray, Dominique. 2006. *The Economics of Knowledge*. Cambridge, MA: MIT Press.

Fortun, Michael. 1998. The Human Genome Project and the Acceleration of Biotechnology. In *Private Science: Biotechnology and the Rise of Molecular Sciences*, ed. Arnold Thackray, 182–201. Philadelphia: University of Pennsylvania Press.

Fortun, Michael. 2001. Mediated Speculations in Genomics Futures Markets. *New Genetics & Society* 20 (2): 139–156.

Fortun, Michael. 2008. *Promising Genomics: Iceland and deCODE Genetics in a World of Speculation*. Berkeley and Los Angeles: University of California Press.

Foucault, Michel. 1975. *Discipline and Punish: The Birth of the Prison*. New York: Random House.

Foucault, Michel. 1984. What Is an Author? In *The Foucault Reader*, ed. Paul Rabinow, 101–120. New York: Pantheon.

FrameNet. n.d. Homepage. https://framenet.icsi.berkeley.edu/fndrupal/. Accessed February 25, 2016.

Franklin, Sarah. 2003. Ethical Biocapital: New Strategies of Cell Culture. In *Remaking Life and Death: Toward an Anthropology of the Biosciences*, ed. Sarah Franklin and Margaret Lock, 97–128. Santa Fe, NM: School of American Research Press.

Fujimura, Joan H. 1987. Constructing "Do-Able" Problems in Cancer Research: Articulating Alignment. *Social Studies of Science* 17 (2): 257–293.

Fujimura, Joan H. 1996. *Crafting Science: A Sociohistory of the Quest for the Genetics of Cancer*. Cambridge, MA: Harvard University Press.

Galison, Peter. 2004. Removing Knowledge. *Critical Inquiry* 31 (1): 229–243.

Galison, Peter, and Emily Ann Thompson, eds. 1999. *The Architecture of Science*. Cambridge, MA: MIT Press.

García-Sancho, Miguel. 2012. *Biology, Computing, and the History of Molecular Sequencing: From Proteins to DNA, 1945–2000*. London: Palgrave Macmillan.

Gaudillière, Jean-Paul, and Hans-Jörg Rheinberger. 2004. *From Molecular Genetics to Genomics: The Mapping Cultures of Twentieth-Century Genetics*. New York: Routledge.

GenBank. 1993. *User Manual and Training Guide and Reference Manual*. Bethesda, MD: GenBank.

Généthon. 1992. *The Généthon Microsatellite Map Catalogue*. Evry, France: Généthon.

Gieryn, Thomas F. 1983. Boundary-Work and the Demarcation of Science from Non-science: Strains and Interests in Professional Ideologies of Scientists. *American Sociological Review* 48 (6): 781–795.

Gieryn, Thomas F. 1999a. *Cultural Boundaries of Science: Credibility on the Line*. Chicago: University of Chicago Press.

Gieryn, Thomas F. 1999b. Two Faces on Science: Building Identities for Molecular Biology and Biotechnology. In *The Architecture of Science*, ed. Peter Galison and Emily Thompson, 423–455. Cambridge, MA: MIT Press.

Gilbert, G. Nigel, and Michael Mulkay. 1984. *Opening Pandora's Box: A Sociological Analysis of Scientists' Discourse*. Cambridge: Cambridge University Press.

Gilbert, Walter. 1987. Creating a New Biology for the Twenty-First Century. *Issues in Science and Technology* 3 (3): 26–35.

Gilbert, Walter. 1991. Towards a New Paradigm Shift in Biology. *Nature* 349 (6305): 99.

Gilbert, Walter. 1992. A Vision of the Grail. In *The Code of Codes: Scientific and Social Issues in the Human Genome Project*, ed. Daniel Kevles and Leroy Hood, 83–97. Cambridge, MA: Harvard University Press.

Gillis, Anria Maria. 1992. The Patent Question of the Year. *Bioscience* 42 (5): 336–339.

Gillis, Justin, and Rick Weiss. 1998. Private Firm Aims to Beat Government to Gene Map. *Washington Post*, May 21.

Glasner, Peter, and Harry Rothman. 2004. *Splicing Life? The New Genetics and Society*. Aldershot, U.K.: Ashgate.

Goad, Walter B. 1983. GenBank. *Los Alamos Science*, Fall, 52–61.

Goffeau, A., B. G. Barrell, H. Bussey, R. W. Davis, B. Dujon, H. Feldmann, F. Galibert, et al. 1996. Life with 6000 Genes. *Science* 274 (5287): 546–567.

Goffman, Erving. 1959. *The Presentation of Self in Everyday Life*. Garden City, NY: Doubleday.

Goffman, Erving. 1974. *Frame Analysis*. New York: Basic Books.

Goffman, Erving. 1981. *Forms of Talk*. Philadelphia: University of Pennsylvania Press.

Gottlieb, Michael, Victoria McGovern, Patricia Goodwin, Stephen Hoffman, and Ayo Oduola. 2000. Please Don't Downgrade the Sequencer's Role ... (letter to the editor). *Nature* 406 (6792): 121–122.

Green, Eric D., Rose M. Mohr, Jacquelyn R. Idol, Myrna Jones, Judy M. Buckingham, Larry L. Deaven, Robert K. Moyzis, and Maynard V. Olson. 1991. Systematic Generation of Sequence-Tagged Sites for Physical Mapping of Human Chromosomes: Application to the Mapping of Human Chromosome 7 Using Yeast Artificial Chromosomes. *Genomics* 11 (3): 548–564.

Green, Philip. 1997. Against a Whole-Genome Shotgun. *Genome Research* 7 (5): 410–417.

Gusfield, Joseph. 1980. *The Culture of Public Problems: Drunk-Driving and the Symbolic Order*. Chicago: University of Chicago Press.

Gusterson, Hugh. 2008. Nuclear Futures: Anticipatory Knowledge, Expert Judgment, and the Lack That Cannot Be Filled. *Science & Public Policy* 35 (8): 551–560.

Guyer, Mark. 1998. Statement on the Rapid Release of Genomic DNA Sequence. *Genome Research* 8 (5): 413.

Hacking, Ian. 1999. *The Social Construction of What?* Cambridge, MA: Harvard University Press.

Hajer, Maarten A. 2009. *Authoritative Governance: Policy-Making in an Age of Mediatization*. Oxford, UK: Oxford University Press.

Harvey, Mark, and Andrew McMeekin. 2007. *Public or Private Economies of Knowledge? Turbulence in the Biological Sciences*. Northamptom, MA: Elgar.

Healy, Bernadine. 1992. Special Report on Gene Patenting. *New England Journal of Medicine* 327 (9): 664–668.

Hedgecoe, Adam. 2004. *The Politics of Personalized Medicine: Pharmacogenetics in the Clinic*. Cambridge: Cambridge University Press.

Hedgecoe, Adam, and Paul Martin. 2003. The Drugs Don't Work: Expectations and the Shaping of Pharmacogenetics. *Social Studies of Science* 33 (3): 327–364.

Helmreich, Stefan. 2008. Species of Biocapital. *Science as Culture* 17 (4): 463–478.

Hess, Charlotte, and Elinor Ostrom. 2003. Ideas, Artifacts, and Facilities: Information as Common-Pool Resource. *Law and Contemporary Problems* 66 (1): 111–146.

Hilgartner, Stephen. 1995. Biomolecular Databases: New Communication Regimes for Biology. *Science Communication* 17 (2): 240–263.

Hilgartner, Stephen. 1998. Data Access Policy in Genome Research. In *Private Science: Biotechnology and the Rise of Molecular Sciences*, ed. Arnold Thackray, 202–218. Philadelphia: University of Pennsylvania Press.

Hilgartner, Stephen. 2000. *Science on Stage: Expert Advice as Public Drama*. Stanford, CA: Stanford University Press.

Hilgartner, Stephen. 2004. Mapping Systems and Moral Order: Constituting Property in Genome Laboratories. In *States of Knowledge: The Co-production of Science and Social Order*, ed. Sheila Jasanoff, 131–141. New York: Routledge.

Hilgartner, Stephen. 2007. Adventurer in Genome Science. *Science* 318 (5854): 1244–1245.

Hilgartner. Stephen. 2009. Intellectual Property and the Politics of Emerging Technology: Inventors, Citizens, and Powers to Shape the Future." *Chicago-Kent Law Review* 84 (1): 197–226.

Hilgartner, Stephen. 2012. Selective Flows of Knowledge in Technoscientific Interaction: Information Control in Genome Research. *British Journal for the History of Science* 45 (2): 267–280.

Hilgartner, Stephen. 2013. Novel Constitutions? New Regimes of Openness in Synthetic Biology. *Biosocieties* 7 (2): 188–207.

Hilgartner, Stephen. 2015. Capturing the Imaginary: Vanguards, Visions, and the Synthetic Biology Revolution. In *Science and Democracy: Making Knowledge and Making Power in the Biosciences and Beyond*, ed. Stephen Hilgartner, Clark Miller, and Rob Hagendijk, 33–55. New York: Routledge.

Hilgartner, Stephen, and Sherry I. Brandt-Rauf. 1994. Data Access, Ownership, and Control: Toward Empirical Studies of Access Practices. *Knowledge: Creation, Diffusion, Utilization* 15 (4): 355–372.

Hilgartner, Stephen, Barbara Prainsack, and J. Benjamin Hurlbut. In press. Ethics as Governance in Genomics and Beyond. In *Handbook of Science and Technology Studies*, ed. *Ulrike Felt, Rayvon Fouché, Clark A. Miller, Laurel Smith-Doerr*, 823–851. Cambridge, MA: MIT Press.

Hine, Christine. 2006. Databases as Scientific Instruments and Their Role in the Ordering of Scientific Work. *Social Studies of Science* 36 (2): 269–298.

Hohfeld, Wesley Newcomb. 1917. Fundamental Legal Conceptions as Applied in Judicial Reasoning. *Yale Law Journal* 26 (8): 710–770.

Holmberg, Tora, and Malin Ideland. 2012. Secrets and Lies: "Selective Openness" in the Apparatus of Animal Experimentation. *Public Understanding of Science* (Bristol, UK) 21 (3): 354–368.

Holtzman, Neil A. 1989. *Proceed with Caution: Predicting Genetic Risks in the Recombinant DNA Era*. Baltimore: Johns Hopkins University Press.

Hood, Leroy. 1992. Biology and Medicine in the Twenty-First Century. In *The Code of Codes: Scientific and Social Issues in the Human Genome Project*, ed. Daniel Kevles and Leroy Hood, 136–163. Cambridge, MA: Harvard University Press.

Hope, Janet. 2008. *Biobazaar: The Open Source Revolution and Biotechnology*. Cambridge, MA: Harvard University Press.

Howlett, Peter, and Mary S. Morgan. 2011. *How Well Do Facts Travel? The Dissemination of Reliable Knowledge*. Cambridge: Cambridge University Press.

Hughes, Thomas P. 1983. *Networks of Power: Electrification in Western Society, 1880–1930*. Baltimore: John Hopkins University Press.

Human Genome Sciences (HGS). 1992. Human Genome Sciences Formed to Develop New Therapeutics from Basic Research on the Human Genome and Makes a $70 Million Grant to the Institute for Genome Research. Press release, July 6.

Human Genome Sciences (HGS). 1994a. *Annual Report to Stockholders*. Rockville, MD: Human Genome Sciences.

Human Genome Sciences (HGS). 1994b. Human Genome Sciences Announces New Agreements with Research Institutions. Press release, July 21.

Human Genome Sciences (HGS). 1994c. Human Genome Sciences to Collaborate with Johns Hopkins Medical School on Colon Cancer Research. Press release, February 28.

Human Genome Organization (HUGO). 1992. Proposed Amendments to the Single Chromosome Workshop Guidelines, edit dates: July 28, August 17, October 1, photocopy.

Human Genome Organization (HUGO). 1995. HUGO Statement on Patenting of DNA Sequences. January.

Human Genome Organization (HUGO). 1996. Summary of Principles Agreed to at the First International Strategy Meeting on Human Genome Sequencing. Bermuda, February 25–28.

Human Genome Organization (HUGO). 1997. Summary of the Report of the Second International Strategy Meeting on Human Genome Sequencing. Bermuda, February 27 to March 2.

Hurlbut, J. Benjamin. 2015. Remembering the Future: Science, Law, and the Legacy of Asilomar. In *Dreamscapes of Modernity: Sociotechnical Imaginaries and the Fabrication of Power*, ed. Sheila Jasanoff and Sang-Hyun Kim, 126–151. Chicago, IL: University of Chicago Press.

Hutchinson, Clyde A. 2007. DNA Sequencing: Bench to Bedside and Beyond. *Nucleic Acids Research* 35 (18): 6227–6237.

Hyman, Richard W. 2001. Sequence Data: Posted vs. Published (letter to the editor). *Science* 291 (5505): 827.

International Human Genome Sequencing Consortium. 2001. Initial Sequencing and Analysis of the Human Genome. *Nature* 409 (6822): 860–921.

Irwin, A. 1998. Gene Scientist Races to Grab a Quick Billion. *Montreal Gazette* , May 15.

Jacob, Marie-Andree, and Annelise Riles, eds. 2007. New Bureaucracies of Virtue. Special issue of *Political and Legal Anthropology Review* 30 (2).

Jaffe, Adam B., and Josh Lerner. 2004. *Innovation and Its Discontents: How Our Broken Patent System Is Endangering Innovation and Progress, and What to Do About It*. Princeton, NJ: Princeton University Press.

Jasanoff, Sheila. 1987. Contested Boundaries in Policy-Relevant Science. *Social Studies of Science* 17 (2): 195–230.

Jasanoff, Sheila. 1990. *The Fifth Branch: Science Advisors as Policymakers*. Cambridge, MA: Harvard University Press.

Jasanoff, Sheila. 1995. *Science at the Bar: Law, Science, and Technology in America*. Cambridge, MA: Harvard University Press.

Jasanoff, Sheila. 1996. Civilization and Madness: The Great BSE Scare of 1996. *Public Understanding of Science* (Bristol, UK) 6:221–232.

Jasanoff, Sheila. 2003. In a Constitutional Moment: Science and Social Order at the Millennium. In *Social Studies of Science and Technology: Looking Back, Ahead; Yearbook of the Sociology of the Sciences*, ed. Bernward Joerges and Helga Nowotny, 155–180. Dordrecht, Netherlands: Kluwer Academic.

Jasanoff, Sheila. 2004a. Ordering Knowledge, Ordering Society. In *States of Knowledge: The Co-production of Science and Social Orders*, ed. Sheila Jasanoff, 13–45. New York: Routledge.

Jasanoff, Sheila, ed. 2004b. *States of Knowledge: The Co-production of Science and Social Orders*. New York: Routledge.

Jasanoff, Sheila. 2005. *Designs on Nature: Science and Democracy in Europe and the United States*. Princeton, NJ: Princeton University Press.

Jasanoff, Sheila, ed. 2012. *Reframing Rights: Bioconstitutionalism in the Genetic Age*. Cambridge, MA: MIT Press.

Jasanoff, Sheila. 2013. Epistemic Subsidiarity—Coexistence, Cosmopolitanism, Constitutionalism. *European Journal of Risk Regulation* 4 (2): 133–141.

Jasanoff, Sheila, and Sang-Hyun Kim, eds. 2015. *Dreamscapes of Modernity: Sociotechnical Imaginaries and the Fabrication of Power*. Chicago: University of Chicago Press.

Joly, Pierre-Benoît. 2015. Governing Emerging Technologies? The Need to Think Outside the (Black) Box. In *Science & Democracy: Making Knowledge and Making Power in the Biosciences and Beyond*, ed. Stephen Hilgartner, Clark Miller, and Rob Hagendijk, 133–155. New York: Routledge.

Joly, Pierre-Benoît, and Vincent Mangematin. 1998. How Long Is Co-operation in Genomics Sustainable? In *The Social Management of Genetic Engineering*, ed. Peter Wheale, Rene von Schomberg and Peter Glasner, 14 pages. Aldershot, UK: Ashgate.

Jordan, Bertrand. 1993. *Travelling around the Human Genome: An in Situ Investigation*. Paris: Editions INSERM; Montrouge: J Libbey Eurotext.

Jordan, Kathleen, and Michael Lynch. 1992. The Sociology of a Genetic Engineering Technique: Ritual and Rationality in the Performance of the "Plasma Prep." In *The Right Tools for the Job: At Work in the Twentieth-Century Life Sciences*, ed. Adele Clarke and Joan H. Fujimura, 77–114. Princeton, NJ: Princeton University Press.

Juengst, Eric T. 1994. Human Genome Research and the Public Interest: Progress Notes from an American Science Policy Experiment. *American Journal of Human Genetics* 54: 121–128.

Juengst, Eric T. 1996. Self-Critical Federal Science? The Ethics Experiment within the Human Genome Project. *Social Philosophy & Policy* 13 (2): 63–95.

Kabat, E. A. 1989. Using Biochemical Data: The Problem with GENBANK. In *Biomolecular Data: A Resource in Transition*, ed. Rita R Colwell, David G. Swartz, Michael Terrell MacDonell, and CODATA, 127–128. Oxford: Oxford University Press.

Kanehisa, M., J. W. Fickett, and W. B. Goad. 1982. A Relational Database System for the Maintenance and Verification of the Los Alamos Sequence Library. *Nucleic Acids Research* 12 (1): 149–158.

Karjala, Dennis. 1992. A Legal Research Agenda for the Human Genome Initiative. *Jurimetrics* 32 (2): 121–222.

Kaufmann, Alain. 2004. Sequencing Human Genomes. In *From Molecular Genetics to Genomics: The Mapping Cultures of Twentieth-Century Genetics*, ed. Jean-Paul Gaudillière and Hans-Jörg Rheinberger, 129–157. New York: Routledge.

Kay, Lily E. 2000. *Who Wrote the Book of Life? A History of the Genetic Code*. Stanford, CA: Stanford University Press.

Keating, Peter, Camille Limoges, and Alberto Cambrosio. 1999. The Automated Laboratory: The Generation and Replication of Work in Molecular Genetics. In *The Practices of Human Genetics; Yearbook of the Sociology of the Sciences*, ed. Michael Fortun and Everett Mendelsohn, 125–142. Boston: Kluwer Academic.

Keller, Evelyn Fox. 2000. *The Century of the Gene*. Cambridge, MA: Harvard University Press.

Kelty, Christopher M. 2008. *Two Bits: The Cultural Significance of Free Software*. Durham, NC: Duke University Press.

Kelty, Christopher M. 2012. This Is Not an Article: Model Organism Newsletters and the Question of "Open Science." *Biosocieties* 7 (2): 140–168.

Kenney, Martin. 1988. *Biotechnology: The University-Industrial Complex*. New Haven, CT: Yale University Press.

Kevles, Daniel J. 1997. Big Science and Big Politics in the United States: Reflections on the Death of the SSC and the Life of the Human Genome Project. *Historical Studies in the Physical and Biological Sciences* 27 (2): 269–297.

Kevles, Daniel J. 1998. *The Baltimore Case: A Trial of Politics, Science, and Character*. New York: Norton.

Kevles, Daniel, and Leroy Hood. 1992. *The Code of Codes: Scientific and Social Issues in the Human Genome Project*. Cambridge, MA: Harvard University Press.

Kieff, F. Scott, ed. 2003. *Perspectives on Properties of the Human Genome Project*. San Diego: Elsevier Academic Press.

Kleinman, Daniel Lee. 2003. *Impure Cultures: University Biology and the World of Commerce*. Madison, WI: University of Wisconsin Press.

Kling, Rob, Lisa B. Spector, and Joanna Fortuna. 2004. The Real Stakes of Virtual Publishing: The Transformation of E-Biomed into PubMed Central. *American Society for Information Science and Technology* 55 (2): 127–148.

Knorr Cetina, Karin. 1981. *The Manufacture of Knowledge: An Essay on the Constructivist and Contextual Nature of Science.* Oxford: Pergamon.

Knorr Cetina, Karin. 1999. *Epistemic Cultures: How the Sciences Make Knowledge.* Cambridge, MA: Harvard University Press.

Kohara, Yugi. 1987. The Physical Map of the Whole *E. coli* Chromosome: Application of a New Strategy for Rapid Analysis and Sorting of a Large Genomic Library. *Cell* 50 (3): 495–508.

Kohler, Robert E. 1994. *Lords of the Fly: Drosophila Genetics and the Experimental Life.* Chicago: University of Chicago Press.

Kopytoff, Igor. 1986. The Cultural Biography of Things: Commoditization as Process. In *The Social Life of Things: Commodities in Cultural Perspective*, ed. Arjun Appadurai, 64–94. Cambridge: Cambridge University Press.

Kowalski, Thomas J. 2000. Analyzing the USPTO's Revised Utility Guidelines. *Nature Biotechnology* 18 (3): 349–350.

Krasner, Stephen D. 1982. Structural Causes and Regime Consequences: Regimes as Intervening Variables. *International Organization* 36 (2): 185–205.

Krimsky, Sheldon. 2003. *Science in the Private Interest: Has the Lure of Private Profits Corrupted Biomedical Research?* Lanham, MD: Rowman & Littlefield.

Kuhn, Thomas S. 1962. *The Structure of Scientific Revolutions.* Chicago: University of Chicago Press.

Lakoff, George, and Mark Johnson. 1980. *Metaphors We Live By.* Chicago: University of Chicago Press.

Latour, Bruno. 1987. *Science in Action: How to Follow Scientists and Engineers through Society.* Cambridge, MA: Harvard University Press.

Latour, Bruno. 1988. *The Pasteurization of France.* Cambridge, MA: Harvard University Press.

Latour, Bruno. 1993. *We Have Never Been Modern.* Cambridge, MA: Harvard University Press.

Latour, Bruno, and Steve Woolgar. 1979. *Laboratory Life: The Construction of Scientific Facts.* Princeton, NJ: Princeton University Press.

Lenoir, Timothy. 1999. Shaping Biomedicine as an Information Science. In *Proceedings of the 1998 Conference on the History and Heritage of Science Information Systems*, ed. Mary Ellen Bowen, Trudi Bellardo Hahn, and Robert V. Williams, 27–45. Medford, NJ: Information Today.

Lenoir, Timothy, and Eric Giannella. 2006. The Emergence and Diffusion of DNA Microarray Technology. *Journal of Biomedical Discovery and Collaboration* 1:1–39.

Leonelli, Sabina. 2011. Packaging Small Facts for Re-use: Databases in Model Organism Biology. In *How Well Do Facts Travel? The Dissemination of Reliable Knowledge*, ed. Peter Howlett and Mary S. Morgan, 325–348. Cambridge: Cambridge University Press.

Leonelli, Sabina. 2012. When Humans Are the Exception: Cross-Species Databases at the Interface of Biological and Clinical Research. *Social Studies of Science* 42 (2): 214–236.

Lessig, Lawrence. 2001. *The Future of Ideas: The Fate of the Commons in a Connected World*. New York: Random House.

Lessig, Lawrence. 2004. *Free Culture: How Big Media Uses Technology and the Law to Lock Down Culture and Control Creativity*. New York: Penguin Press.

Lewenstein, Bruce V. 1995. From Fax to Facts: Communication in the Cold Fusion Saga. *Social Studies of Science* 25 (3): 403–436.

Lewontin, Richard. 1991. *Biology as Ideology: The Doctrine of DNA*. New York: Harper Collins.

Little, Peter. 1991. Complementary Questions. *Nature* 352 (6330): 20–21.

Löwy, Ilana. 1988. Ludwik Fleck on the Social Construction of Medical Knowledge. *Sociology of Health & Illness* 10 (2): 133–155.

Löwy, Ilana, and Jean Paul Gaudillière. 2008. Localizing the Global: Testing for Hereditary Risks of Breast Cancer. *Science, Technology & Human Values* 33 (3): 299–325.

Lynch, Michael. 1985. *Art and Artifact in Laboratory Science: A Study of Shop Work and Shop Talk in a Research Laboratory*. New York and London: Routledge.

Lynch, Michael. 1993. *Scientific Practice and Ordinary Action: Ethnomethodology and Social Studies of Science*. New York: Cambridge University Press.

Lynch, Michael, Simon A. Cole, Ruth McNally, and Kathleen Jordan. 2008. *Truth Machine: The Contentious History of DNA Fingerprinting*. Chicago: University of Chicago Press.

Lynch, Michael, Stephen Hilgartner, and Carin Berkowitz. 2005. Voting Machinery, Counting, and Public Proofs in the 2000 US Presidential Election. In *Making Things Public: Atmospheres of Democracy*, ed. Bruno Latour and Peter Weibel, 814–825. Cambridge, MA: MIT Press.

Macilwain, Colin. 2000. Biologists Challenge Sequencers on Parasite Genome Publication. *Nature* 405 (6787): 601–602.

Maddox, John. 1989a. Making Authors Toe the Line. *Nature* 342 (6252): 855.

Maddox, John. 1989b. Making Good Databanks Better. *Nature* 341 (6240): 277.

Maier, E., J. D. Hoheisel, L. McCarthy, R. Mott, A. V. Grigoriev, A. P. Monaco, Z. Larin, and H. Lehrach. 1992. Complete Coverage of the *Schizosaccharomyces pombe* Genome in Yeast Artificial Chromosomes. *Nature Genetics* 1 (4): 273–277.

Marshall, Eliot. 1994a. The Company That Genome Researchers Love to Hate. *Science* 266 (5192): 1800–1802.

Marshall, Eliot. 1994b. A Showdown over Gene Fragments. *Science* 266 (5183): 208–210.

Marshall, Eliot. 1995. Human Genome Project: Emphasis Changes from Mapping to Large Scale Sequencing. *Science* 268:1270–1271.

Marshall, Eliot. 1996. Genome Sequences Take the Pledge. *Science* 272 (5621): 477–478.

Mathews, Jessica. 1991. The Race to Claim the Gene. *Washington Post*, November 17.

Maxam, Allan, and Walter Gilbert. 1977. A New Method for Sequencing DNA. *Proceedings of the National Academy of Sciences* 74 (2): 560–564.

McCain, Katherine W. 1995. Mandating Sharing: Journal Policies in the Natural Sciences. *Science Communication* 16 (4): 403–431.

McCray, W. Patrick. 2012. *The Visioneers: How a Group of Elite Scientists Pursued Space Colonies, Nanotechnologies, and a Limitless Future.* Princeton, NJ: Princeton University Press.

McElheny, Victor K. 2010. *Drawing the Map of Life: Inside the Human Genome Project.* New York: Basic Books.

McKusick, Victor A., and Frank H. Ruddle. 1987. A New Discipline, a New Name, a New Journal. *Genomics* 1:1–2.

McNally, Ruth, and Peter Glasner. 2007. Survival of the Gene? Twenty-First Century Visions from Genomics, Proteomics, and the New Biology. In *New Genetics, New Social Formations*, ed. Paul Atkinson, Peter Glasner, and Helen Greenslade, 253–278. New York: Routledge.

Merck & Company, Inc. 1995. Merck Gene Index. Press release, February 10.

Merton, Robert K. 1973. *The Sociology of Science: Theoretical and Empirical Investigations.* Chicago: University of Chicago Press.

Metlay, Grischa. 2006. Reconsidering Renormalization: Stability and Change in 20th-Century Views on University Patents. *Social Studies of Science* 36 (4): 565–597.

Mewes, H. W., R. Doelz, and D. G. George. 1994. Sequence Databases: An Indispensible Source for Biotechnological Research. *Journal of Biotechnology* 35 (2): 239–256.

Minsky, Marvin. [1974] 1975. A Framework for Representing Knowledge." MIT Artificial Intelligence Laboratory Memo 306, June 1974. Reprinted in *The Psychology of Computer Vision*, ed. Patrick Henry Winston, 211–277. New York: McGraw-Hill.

Mirowski, Philip. 2011. *Science-Mart: Privatizing American Science*. Cambridge, MA: Harvard University Press.

Mol, Annemarie. 2002. *The Body Multiple: Ontology in Medical Practice*. Durham, NC: Duke University Press.

Moore, Stephen D. 1993. SmithKline Maps Out Research Route—Drug Maker on New Road with Gene Data Company. *Wall Street Journal*, September 7.

Moss, Lenny. 2004. *What Genes Can't Do*. Cambridge, MA: MIT Press.

Mulkay, Michael. 1976. Norms and Ideology in Science. *Social Sciences Information / Information sur les Sciences Sociales* 15 (4–5): 637–656.

Murphy, Michelle. 2006. *Sick Building Syndrome and the Problem of Uncertainty: Environmental Politics, Technoscience, and Women Workers*. Durham, NC: Duke University Press.

Myers, Eugene W., Granger G. Sutton, Hamilton O. Smith, Mark D. Adams, and J. Craig Venter. 2002. On the Sequencing and Assembly of the Human Genome. *Proceedings of the National Academy of Sciences of the United States of America* 99 (2): 4145–4156.

Myers, Greg. 1990. *Writing Biology: Texts in the Social Construction of Biology*. Madison: University of Wisconsin Press.

National Center for Human Genome Research (NCHGR). 1989. Human Genome Program Center Grants (P30, P50). *NIH Guide* 18 (October 13): 8–12.

National Center for Human Genome Research (NCHGR). 1992a. Briefing book for Meeting of the NIH–DOE Joint Subcommittee and the Eighth Meeting of the NIH Program Advisory Committee on the Human Genome, December 7 and 8, Bethesda, MD, photocopy.

National Center for Human Genome Research (NCHGR). 1992b. *Genome Report Card–March 1992*. Bethesda, MD: NCHGR.

National Center for Human Genome Research (NCHGR). 1995. Pilot Projects for Sequencing of the Human Genome. *NIH Guide* 24 (March 10). https://grants.nih.gov/grants/guide/rfa-files/RFA-HG-95-005.html. Accessed April 15, 2016.

National Human Genome Research Institute (NHGRI). 1998. Large-Scale Sequencing Workshop Report. September 3–4. https://www.genome.gov/10001489/largescale-sequencing-workshop/. Accessed April 15, 2016.

National Human Genome Research Institute (NHGRI). 2000. NHGRI Policy for Release and Database Deposition of Sequence Data. https://www.genome .gov/10000910/policy-on-release-of-human-genomic-sequence-data-2000/. Accessed April 15, 2016.

National Human Genome Research Institute (NHGRI). 2003. Reaffirmation and Extension of NHGRI Rapid Data Release Policies: Large-Scale Sequencing and Other Community Resource Projects. February. https://www.genome.gov/10506537/ reaffirmation-and-extension-of-nhgri-rapid-data-release-policies/. Accessed April 15, 2016.

National Institutes of Health (NIH). 1992. Statement by Dr. Bernadine Healy, Director, National Institutes of Health. Press release, July 6. Bethesda, MD: National Institutes of Health.

National Institutes of Health (NIH) Ad Hoc Committee. 1990. "Ad Hoc Committee on the Role of cDNAs in the Human Genome Program: Meeting Summary, December 2, 1990, Bethesda, MD," photocopy.

National Institutes of Health (NIH) and US Department of Energy (DOE). 1990. *Understanding Our Genetic Inheritance. The Human Genome Project: The First Five Years, FY 1991–1995*. Springfield, VA: National Technical Information Service.

National Research Council (NRC). 1988. *Mapping and Sequencing the Human Genome*. Washington, DC: National Academies Press.

National Research Council (NRC). 1997. *Intellectual Property and Research Tools in Molecular Biology*. Washington, DC: National Academies Press.

National Research Council (NRC). 2003a. *Large-Scale Biomedical Science: Exploring Strategies for Future Research*. Washington, DC: National Academies Press.

National Research Council (NRC). 2003b. *Sharing Publication-Related Data and Materials: Responsibilities of Authorship in the Life Sciences*. Washington, DC: National Academics Press.

NCBI News (National Center for Biotechnology Information). 1995. "Bank-It." March. Pages 1–2.

Nelkin, Dorothy. 1987. *Selling Science: How the Press Covers Science and Technology*. New York: W.H. Freeman.

Nelkin, Dorothy, and Laurence R. Tancredi. 1989. *Dangerous Diagnostics: The Social Power of Biological Information*. Chicago: University of Chicago Press.

Nelson, Nicole. 2013. Modeling Mouse, Human, and Discipline: Epistemic Scaffolds in Animal Behavior Genetics. *Social Studies of Science* 43 (1): 3–29.

Nowak, Rachel. 1994. NIH in Danger of Losing Out on *BRCA1* Patent. *Science* 266 (5183): 209.

Nukaga, Yoshio, and Alberto Cambrosio. 1997. Medical Pedigrees and the Visual Production of Family Disease in Canadian and Japanese Genetic Counselling Practice. *Sociology of Health & Illness* 19 (19B): 29–55.

Office of Human Genome Research. 1989. Human Genome Program Center Grants (P30). *NIH Guide* 18 (July 21): 7–10.

Office of Technology Assessment. 1988. *Mapping Our Genes—Genome Projects: How Big? How Fast?* OTA-BA-373. Washington, DC: US Government Printing Office.

Olson, Maynard V. 1995. A Time to Sequence. *Science* 270 (5235): 394–396.

Olson, Maynard V. 2002. The Human Genome Project: A Player's Perspective. *Journal of Molecular Biology* 319:931–942.

Olson, Maynard V., Leroy Hood, Charles Cantor, and David Botstein. 1989. A Common Language for Physical Mapping of the Human Genome. *Science* 245 (4925): 1434–1435.

Orwell, George. 1949. *Nineteen Eighty-Four.* London: Secker and Warburg.

Ostrom, Elinor. 1990. *Governing the Commons: The Evolution of Institutions for Collective Action.* Cambridge: Cambridge University Press.

Ouellette, B. F. Francis, and Mark S. Boguski. 1997. Database Divisions and Homology Search Files: A Guide for the Perplexed. *Genome Research* 7 (10): 952–955.

Pace, Tomasino. 2001. When Public-Interest Science Needs Solidarity. *Nature* 406 (6792): 122.

Papadopoulos, N., Nicholas C. Nicolaides, Ying-Fei Wei, Steven M. Ruben, Kenneth C. Carter, Craig R. Rosen, William A. Haseltine, Robert D. Fleischmann, Claire M. Fraser, Mark D. Adams, J. Craig Venter, Stanley R. Hamilton, Gloria M. Petersen, Patrice Watson, Henry T. Lynch, Päivi Peltomäki, Jukka-Pekka Mecklin, Albert de la Chapelle, Kenneth W. Kinzler, and Bert Vogelstein. 1994. Mutation of a *mutL* Homolog in Hereditary Colon Cancer *Science* 263 (5153): 1625–1629.

Parthasarathy, Shobita. 2007. *Building Genetic Medicine: Breast Cancer, Technology, and the Comparative Politics of Health Care.* Cambridge, MA: MIT Press.

Parthasarathy, Shobita. 2011. Whose Knowledge? What Values? The Comparative Politics of Patenting Life Forms in the United States and Europe. *Policy Sciences* 44 (3): 267–288.

Peñalver, Eduardo M., and Sonia K. Katyal. 2010. *Property Outlaws: How Squatters, Pirates, and Protesters Improve the Law of Ownership.* New Haven, CT: Yale University Press.

Perkin-Elmer. 1998a. Celera Is Name of New Genomics Company Formed by Perkin-Elmer and Dr. J. Craig Venter. Press release, August 5.

Perkin-Elmer. 1998b. Perkin-Elmer to Introduce New Instrument Based on Breakthrough DNA Analysis. Press release, May 9.

Perkin-Elmer and The Institute for Genome Research (TIGR). 1998. Perkin-Elmer, Craig Venter, and TIGR Announce Formation of New Genomics Company; Plan to Sequence Human Genome within Three Years. Press release, May 9.

Pinch, Trevor. 1981. The Sun-Set: The Presentation of Certainty in Scientific Life. *Social Studies of Science* 11 (1): 131–158.

Pinch, Trevor. 1986. *Confronting Nature: The Sociology of Solar-Neutrino Detection.* Dordrecht, Netherlands: Reidel.

Powell, Walter W., and Paul DiMaggio, eds. 1991. *The New Institutionalism in Organizational Analysis.* Chicago: University of Chicago Press.

Proctor, Robert, and Londa L. Schiebinger. 2008. *Agnotology: The Making and Unmaking of Ignorance.* Stanford, CA: Stanford University Press.

Rabeharisoa, Vololona, and Michel Callon. 2004. Patients and Scientists in French Muscular Dystrophy Research. In *States of Knowledge: The Co-production of Science and Social Order,* ed. Sheila Jasanoff, 142–160. New York: Routledge.

Rabinow, Paul. 1996. *Making PCR: A Story of Biotechnology.* Chicago: University of Chicago Press.

Rabinow. Paul. 1999. *French DNA: Trouble in Purgatory.* Chicago: University of Chicago Press.

Rabinow, Paul, and Gaymon Bennett. 2012. *Designing Human Practices: An Experiment with Synthetic Biology.* Chicago: University of Chicago Press.

Rabinow, Paul, and Talia Dan-Cohen. 2005. *A Machine to Make the Future: Biotech Chronicles.* Princeton, NJ: Princeton University Press.

Rader, Karen. 2004. *Making Mice: Standardizing Animals for American Biomedical Research, 1900–1955.* Princeton, NJ: Princeton University Press.

Recer, P. 1998. Gene Technology Must Be Proven Before Government Will Use It. Associated Press, May 11.

Rheinberger, Hans-Jörg. 1997. *Toward a History of Epistemic Things: Synthesizing Proteins in the Test Tube.* Stanford, CA: Stanford University Press.

Richardson, Sarah S., and Hallam Stevens, eds. 2015. *Postgenomics: Perspectives on Biology after the Genome.* Durham, NC: Duke University Press.

Roberts, Leslie. 1987. Who Owns the Human Genome? *Science* 237 (4813): 358–361.

Roberts, Leslie. 1989a. New Game Plan for Genome Mapping. *Science* 245 (4925): 1438–1440.

Roberts, Leslie. 1989b. Plan for Genome Centers Sparks a Controversy. *Science* 246 (4927): 204–205.

Roberts, Leslie. 1991. Gambling on a "Shortcut" to Genome Sequencing. *Science* 252 (5013): 1618–1619.

Roberts, Leslie. 1992. NIH Gene Patents, Round Two. *Science* 255 (5047): 912–913.

Roberts, Richard J. 1989. Benefits of Databases (letter to the editor). *Nature* 342 (6246): 114.

Rödder, Simone. 2009. Reassessing the Concept of a Medialization of Science: A Story from the "Book of Life." *Public Understanding of Science* (Bristol, UK) 18 (4): 452–463.

Rödder, Simone, Martina Franzen, and Peter Weingart, eds. 2012. *The Sciences' Media Connection—Public Communication and Its Repercussions*. Dordrecht, Netherlands: Springer.

Rommens, J. M., M. C. Iannuzzi, B. Kerem, M. L. Drumm, G. Melmer, M. Dean, R. Rozmahel, et al. 1989. Identification of the Cystic Fibrosis Gene: Chromosome Walking and Jumping. *Science* 245 (4922): 1059–1065.

Roos, David S. 2001. Computational Biology: Bioinformatics—Trying to Swim in a Sea of Data. *Science* 291 (5507): 1260–1261.

Rowen, Lee. 2007. Sequencing the Human Genome: A Historical Perspective on Challenges for Systems Integration. In *BioMEMS and Biomedical Nanotechnology*, vol. 2: *Micro/Nano Technologies for Genomics and Proteomics*, ed. Mauro Ferrari, Mihrimah Ozkan, and Michael J. Heller, 365–99. Boston: Springer.

Rowen, Lee, and Leroy Hood. 2000. Response: Publication Rights for Sequence Data Producers (letter to the editor). *Science* 290 (5497): 1696–1698.

Rowen, Lee, Gane K. S. Wong, Robert P. Lane, and Leroy Hood. 2000. Publication Rights in the Era of Open Data Release Policies. *Science* 289 (5486): 1881.

Schäfer, Mike S. 2009. From Public Understanding to Public Engagement: An Empirical Assessment of Changes in Science Coverage. *Science Communication* 30 (4): 475–505.

Service, Roberts J. 1994. Stalking the Start of Colon Cancer. *Science* 263 (5153): 1559–1560.

Shapin, Steven. 1988. The House of Experiment in Seventeenth-Century England. *Isis* 79 (3): 373–404.

Shapin, Steven. 1996. *The Scientific Revolution*. Chicago: University of Chicago Press.

Shapin, Steven. 2008. *The Scientific Life: A Moral History of a Late Modern Vocation*. Chicago: University of Chicago Press.

Shapin, Steven, and Simon Schaffer. 1985. *Leviathan and the Air-Pump: Hobbes, Boyle, and the Experimental Life*. Princeton, NJ: Princeton University Press.

Sharing Data from Large-Scale Biological Research Projects: A System of Tripartite Responsibility (Fort Lauderdale Agreement). 2003. Report of a meeting organized by the Wellcome Trust and held on January 14–15, 2003, at Fort Lauderdale, Florida. https://www.genome.gov/pages/research/wellcomereport0303.pdf.

Shreeve, James. 2004. *The Genome War: How Venture Tried to Capture the Code of Life and Save the World*. New York: Knopf.

Shrum, Wesley, Joel Genuth, and Ivan Chompalov. 2007. *Structures of Scientific Collaboration*. Cambridge, MA: MIT Press.

Skinner, Quentin. 1969. Meaning and Understanding in the History of Ideas. *History and Theory* 8 (1): 3–53.

Slaughter, Sheila, and Gary Rhoads. 2004. *Academic Capitalism and the New Economy: Markets, State, and Higher Education*. Baltimore: Johns Hopkins University Press.

Smith, Cassandra L., Jason G. Econome, Andrew Schutt, Stephanie Klco, and Charles R. Cantor. 1987. A Physical Map of the *Escherichia coli* K12 Genome. *Science* 236 (4807): 1448–1453.

Smith, L. M., J. Z. Sanders, R. J. Kaiser, P. Hughes, C. Dodd, C. R. Connell, C. Heiner, S. B. Kent, and L. E. Hood. 1986. Fluorescence Detection in Automated DNA Sequence Analysis. *Nature* 321 (6071): 674–679.

Smith, Lloyd M., Leroy E. Hood, Michael W. Hunkapiller, and Tim J. Hunkapiller. 1992. *Automated DNA Sequencing Technique*. U.S. Patent No. 5,171,534, filed October 15, 1990, and issued December 15, 1992.

Stallman, Richard M. 2010. *Free Software, Free Society: Selected Essays of Richard M. Stallman*. 2nd ed. Boston: GNU Press.

Star, Susan Leigh. 1999. The Ethnography of Infrastructure. *American Behavioral Scientist* 43 (3): 377–391.

Star, Susan Leigh, and Karen Ruhleder. 1996. Steps toward an Ecology of Infrastructure: Design and Access for Large Information Spaces. *Information Systems Research* 7 (1): 111–134.

Stemerding, Dirk, and Stephen Hilgartner. 1998. Means of Co-ordination in the Making of Biological Science: On the Mapping of Plants, Animals, and Genes. In *Getting New Technologies Together: Studies in Making Sociotechnical Order*, ed. Cornelius Disco and Barend van der Meulen, 39–69. Boston: DeGruyter.

Stevens, Hallam. 2013. *Life Out of Sequence: A Data-Driven History of Bioinformatics*. Chicago: University of Chicago Press.

Stout, Hilary. 1992. Gene-Fragment Patent Request Is Turned Down. *Wall Street Journal* September 23, B1, B7.

Strasser, Bruno J. 2008. GenBank—Natural History in the 21st Century. *Science* 322 (5901): 537–539.

Strasser, Bruno J. 2010. Collecting, Comparing, and Computing Sequences: The Making of Margaret O. Dayhoff's "Atlas of Protein Sequence and Structure," 1954–1965. *Journal of the History of Biology* 43 (4): 623–660.

Strasser, Bruno J. 2011. The Experimenter's Museum: GenBank, Natural History, and the Moral Economies of Biomedicine. *Isis* 102 (1): 60–96.

Sulston, John, and Georgina Ferry. 2002. *The Common Thread: A Story of Science, Politics, Ethics, and the Human Genome.* Washington, DC: Joseph Henry Press.

Sunder Rajan, Kaushik. 2006. *Biocapital: The Constitution of Postgenomic Life.* Durham, NC: Duke University Press.

Terry, Sharon F. 2003. Learning Genetics. *Health Affairs* 22 (5): 166–171.

Thackray, Arnold. 1998. *Private Science: Biotechnology and the Rise of Molecular Sciences.* Philadelphia: University of Pennsylvania Press.

The Institute for Genome Research (TIGR). 1992. The Institute for Genome Research Formed to Accelerate Gene Discovery from the Human Genome. Press release, July 6.

The Institute for Genome Research (TIGR). 1997. TIGR/HGS Funding Relationship Reaches Early Conclusion, Press release, June 24.

Tilly, Charles. 1978. *From Mobilization to Revolution.* New York: McGraw Hill.

Tilly, Charles, Sidney Tarrow, and Doug McAdam. 2006. *Dynamics of Contention.* New York: Cambridge University Press.

Traweek, Sharon. 1988. *Beamtimes and Lifetimes: The World of High Energy Physicists.* Cambridge, MA: Harvard University Press.

Turner, Fred. 2006. *From Counterculture to Cyberculture: Stewart Brand, the Whole Earth Network, and the Rise of Digital Utopianism.* Chicago: University of Chicago Press.

Tutton, Richard. 2011. Promising Pessimism: Reading the Futures to Be Avoided in Biotech. *Social Studies of Science* 41 (3): 411–429.

US House of Representatives, Committee on Energy and the Environment, Subcommittee on Energy and Environment. 1998. *Hearing on Human Genome Project: How Private Sector Developments Affect the Government Program.* 105th Cong., Second sess., June 17.

Van Lente, Harro, and Arie Rip. 1998. The Rise of Membrane Technology: From Rhetorics to Social Reality. *Social Studies of Science* 28 (2): 221–254.

Venter, J. Craig. 2007. *A Life Decoded: My Genome, My Life*. New York: Viking.

Venter, J. Craig, Mark D. Adams, Eugene W. Myers, Peter W. Li, Richard J. Mural, Granger G. Sutton, Hamilton O. Smith, et al. 2001. The Sequence of the Human Genome. *Science* 291 (5507): 1304–1351.

Vogel, Kathleen M. 2013. *Phantom Menace or Looming Danger? A New Framework for Assessing Bioweapons Threats*. Baltimore: Johns Hopkins University Press.

Vollrath, D., S. Foote, A. Hilton, L. Brown, P. Beer-Romero, J. Bogan, and D. Page. 1992. The Human Y Chromosome: A 43-Interval Map Based on Naturally Occurring Deletions. *Science* 258 (5079): 52–59.

Wade, Nicholas. 1998a. Beyond Sequencing of Human DNA. *New York Times*, May 12.

Wade, Nicholas. 1998b. International Gene Project Gets Lift. *New York Times*, May 17.

Wade, Nicholas. 1998c. Scientist's Plan: Map All DNA within Three Years. *New York Times*, May 10.

Waldholtz, Michael. 1994. *Medicine & Health (A Special Report): What Ails Us—Tracing Tumors: The Search for 'Cancer Genes' May Help Solve One of the Greatest Medical Mysteries of the Century. Wall Street Journal*, March 20.

Walker, Richard T. 1989. A Method for the Rapid and Accurate Deposition of Nucleic Acid Data in an Acceptably Annotated Form. In *Biomolecular Data: A Resource in Transition*, ed. Rita R Colwell, David G. Swartz, Michael Terrell MacDonell, and CODATA, 45–52. Oxford: Oxford University Press.

Wall Street Journal. 1993. SmithKline Pact with HGS. May 21.

Waterston, Robert H., Eric S. Lander, and John E. Sulston. 2002. On the Sequencing of the Human Genome. *Proceedings of the National Academy of Sciences of the United States of America* 99 (6): 3712–3716.

Waterston, Robert H., Eric S. Lander, and John E. Sulston. 2003. More on the Sequencing of the Human Genome. *Proceedings of the National Academy of Sciences of the United States of America* 100 (6): 3022–3024.

Waterston, Robert H., and John E. Sulston. 1998. The Human Genome Project: Reaching the Finish Line. *Science* 282 (5386): 53–54.

Watson, James D. 1990. The Human Genome Project: Past, Present, and Future. *Science* 248 (4951): 44–49.

Weber, James L., and Eugene W. Myers. 1997. Human Whole-Genome Shotgun Sequencing. *Genome Research* 7 (5): 401–409.

Wellcome Trust. 1998. Wellcome Trust Announces Major Increase in Human Genome Sequencing. Press release, May 13.

Wells, Robert D. 1989. To Publish or Not to Publish DNA Sequences. In *Biomolecular Data: A Resource in Transition*, ed. Rita R. Colwell, David G. Swartz, Michael Terrell MacDonell, and CODATA, 335–338. Oxford: Oxford University Press.

Wheeler, David L. 1991. Scientists Fear NIH Attempt to Patent Genes Will Threaten Collaboration on Research. *Chronicle of Higher Education* October 21: A27–A28.

Wheeler, David L. 1992a. NIH Files Patent Application on 2,375 More Human Genes. *Chronicle of Higher Education* February 19: A23.

Wheeler, David L. 1992b. Using Powerful Machines, an NIH Researcher Leads Effort to Identify Human Genes. *Chronicle of Higher Education* February 26: A7–A8.

Whitehead Institute/MIT Center for Genome Research. 1995. Human Genomic Mapping Project, Data Release 5. January, Cambridge, Mass.

William Blair & Company. 1999. *PE Corporation—Celera Genomics Group* (Basic Report: Winton Gibbons, December 9). Chicago: William Blair & Company.

Wilcox, Andrea S., Akbar S. Khan, Janet A. Hopkins, and James M. Sikela. 1991. Use of 3′ Untranslated Sequences of Human cDNAs for Rapid Chromsome Assignment and Conversion to STSs: Implications for an Expression Map of the Genome. *Nucleic Acids Research* 19 (8): 1837–1843.

Winickoff, David E. 2015. Biology Denatured: The Public-Private Lives of Lively Things. In *Science and Democracy: Making Knowledge and Making Power in the Biosciences and Beyond*, ed. Stephen Hilgartner, Clark Miller, and Rob Hagendijk, 15–32. New York: Routledge.

Woolgar, Steve, and Javier Lezaun. 2013. The Wrong Bin Bag: A Turn to Ontology in Science and Technology Studies? *Social Studies of Science* 43 (3): 321–340.

Zehetner, Günther, ed. 1991. *Reference Library News Update* [photocopied newsletter] 1 (May): 1–21.

Zweiger, Gary. 2001. *Transducing the Genome: Information, Anarchy, and Revolution in the Biomedical Sciences*. New York: McGraw Hill.

Index

Inside Technology

edited by Wiebe E. Bijker, W. Bernard Carlson, and Trevor Pinch

http://mitpress.mit.edu/books/series/inside-technology